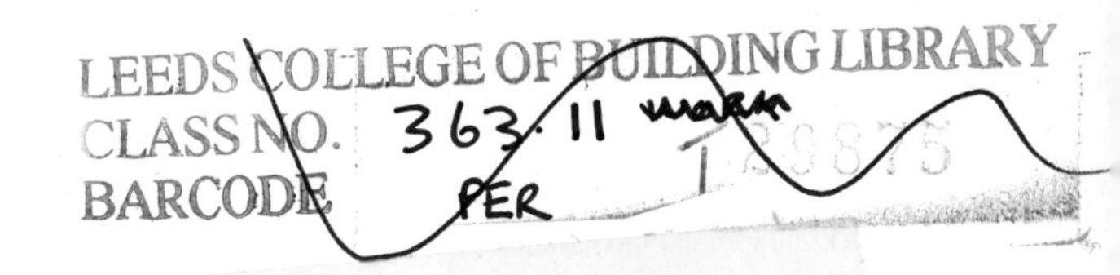

Health and safety questions and answers: a practical approach

Pat Perry

Thomas Telford

Published by Thomas Telford Publishing, Thomas Telford Ltd, 1 Heron Quay, London E14 4JD.
URL: http://www.thomastelford.com

Distributors for Thomas Telford books are
USA: ASCE Press, 1801 Alexander Bell Drive, Reston, VA 20191-4400, USA
Japan: Maruzen Co. Ltd, Book Department, 3–10 Nihonbashi 2-chome, Chuo-ku, Tokyo 103
Australia: DA Books and Journals, 648 Whitehorse Road, Mitcham 3132, Victoria

First published 2003

Also in this series from Thomas Telford Books
Construction safety: questions and answers. Pat Perry. ISBN 0 7277 3233 1
Fire safety: questions and answers. Pat Perry. ISBN 07277 3239 0
Risk assessment: questions and answers. Pat Perry. ISBN 07277 3238 2
CDM questions and answers: a practical approach 2nd edition. Pat Perry. ISBN 07277 3107 6

A catalogue record for this book is available from the British Library

ISBN: 0 7277 3240 4

Any safety sign or symbol used in this book is for illustrative purposes only and does not necessarily imply that the sign or symbol used meets any legal requirements or good practice guides. Before producing any sign or symbol, the reader is recommended to check with the relevant British Standard or the Health and Safety (Safety Signs and Signals) Regulations 1996.

Throughout the book the personal pronouns 'he', 'his', etc. are used when referring to 'the Client', 'the Designer', 'the Planning Supervisor', etc., for reasons of readability. Clearly, it is quite possible these hypothetical characters may be female in 'real-life' situations, so readers should consider these pronouns to be grammatically neuter in gender, rather than masculine.

Typeset by Alex Lazarou, Surbiton, Surrey
Printed and bound in Great Britain by MPG Books, Bodmin, Cornwall

Biography

Pat Perry, MCIEH, MIOSH, FRSH, MIIRM, qualified as an Environmental Health Officer in 1978 and spent the first years of her career in local government enforcing environmental health laws, in particular health and safety law, which became her passion. She has extensive knowledge of her subject and has served on various working parties on both health and safety and food safety. Pat contributes regularly to professional journals, e.g. *Facilities Business,* and has been commissioned by Thomas Telford Publishing to write a series of health and safety books.

After a period in the private sector, Pat set up her own environmental health consultancy, Perry Scott Nash Associates Ltd, in the latter part of 1988, and fulfilled her vision of a 'one-stop shop' for the provision of consultancy services to the commercial and retail sectors.

The consultancy has grown considerably over the years and provides consultancy advice to a wide range of clients in a variety of market sectors. Leisure and retail have become the consultancy's major expertise and the role of planning supervisor and environmental health consultant is provided on projects ranging from a few hundred thousand pounds to many millions, e.g. new public house developments and major department store refits and refurbishments.

Perry Scott Nash Associates Ltd have strong links to the enforcing agencies; consultants having come mostly from similar backgrounds and approach projects and all the issues and concerns associated with legal compliance with pragmatism and commercial understanding.

Should you wish to contact Pat Perry about any issue in this book, or to enquire further about the consultancy services offered by Perry Scott Nash Associates Ltd, please contact us direct at:

Perry Scott Nash Associates Ltd
Perry Scott Nash House
Primett Road
Stevenage
Herts
SG1 3EE

Alternatively phone, fax or email on:

Tel: 01438 745771
Fax: 01438 745772
Email: p.perry@perryscottnash.co.uk

We would also recommend that you visit our website at:
www.perryscottnash.co.uk

Acknowledgements

My sincere thanks go to Maureen for her never ending support and encouragement and to Janine and the Business Support team at Perry Scott Nash Associates Ltd for typing all the handwritten manuscripts with such patience and efficiency.

Author's note

Many of the publications referenced in this book are available for download on a number of websites, e.g.:

- www.hse.gov.uk/pubns/index.htm
- www.hsebooks.co.uk

Also, guidance on the availability of books is available from the HSE Info Line on 0541 545 500.

Contents

1

Introduction to health and safety

What is health and safety all about?

Health and safety is all about keeping those employed in your business and those affected by the things that you do safe and free from ill-health.

For many decades people have been killed or seriously injured during the course of their employment because they were required to work in unsafe and unhealthy working conditions.

Employers and others have legal duties to ensure the 'health, safety and welfare' of employees and others while they are at work or resorting to the premises or 'undertaking'.

Health and safety started during the Industrial Revolution; it is not something new. But, as our knowledge increases so does the expectation that measures to protect people will also improve.

Does health and safety law only apply to employers?

No. While employers carry the major share of responsibility for ensuring that standards of health and safety are maintained, there are other people who also have responsibilities for ensuring the health, safety and welfare of employees and others, namely:

- persons in control of premises
- self-employed persons
- building/property owners
- 'duty holders' as defined in some Regulations.

Depending on the circumstances, legal responsibilities for health and safety compliance will rest with those people, in addition to, or instead of, the employer.

Isn't all the UK's health and safety legislation a result of EU directives and 'interference'?

No. Health and safety law existed long before Britain joined the EU and laws such as:

- The Factories Act 1961
- The Offices Shops and Railway Premises Act 1963

were in place and enforced by various statutory bodies.

When Britain joined the European Union it accepted that certain 'social' legislation would need to be introduced across Europe to harmonise individual national laws so as to create a 'level playing field' for the purposes of commercial competition.

It is certainly true to say that the UK's health and safety law is now heavily influenced by EU requirements and there is a constant stream of health and safety regulatory revision and the introduction of new, or revised, regulations and statutes.

Again, as our knowledge increases, e.g. due to scientific research, the expectation from hazard and risk becomes commonplace. Older legislative requirements no longer fit the bill – they do not protect people from modern-day hazards and risks and so the expected control measures have to be increased — hence the need for new laws.

The UK has, since the introduction of the Health and Safety at Work Etc. Act 1974, had an approach to health and safety which has been based on a broadly principled self-regulatory approach. The EU approach to health and safety, however, is to be more prescriptive, i.e. it states what will be done as opposed to setting out an objective to be achieved.

As an employer, what am I responsible for in respect of health and safety?

As an employer, you are responsible for the 'health, safety and welfare' of your employees while at work, and while they work away from your usual premises, and you are also responsible for the health, safety and welfare of people who 'resort' to your premises, e.g. contractors, visitors, tradesmen, delivery drivers, etc. and also for those who may be affected by your business or undertaking, e.g. public, customers, passers-by, children, trespassers, etc.

The responsibilities an employer carries mean that you have to:

- provide a safe place of work
- maintain the place of work in a safe condition
- provide safe means of access to and egress from the place of work
- maintain plant and equipment in a safe condition
- prevent the exposure of your employees and others to unacceptable hazards which could cause injury or ill-health
- ensure that employees follow a safe system of work
- ensure that employees receive information, instruction and training
- ensure that you prepare a written statement of policy in respect of health and safety if you have five or more employees
- carry out Risk Assessments of work activities and record the significant findings in writing if you have five or more employees
- carry out assessments in respect of the use of hazardous, dangerous and explosive substances

- carry out Fire Risk Assessments
- use and consult with competent persons
- prepare emergency plans and procedures
- consult with and inform your staff on health and safety issues.

Are there any other responsibilities that employers have in respect of health and safety?

Yes, there is an additional major responsibility and that is to have current Employers' Liability insurance cover.

If employees are injured while they are at work, they may have the right to claim compensation from you if they believe you are responsible. The Employers' Liability (Compulsory Insurance) Act 1969 ensures that you have at least a minimum level of insurance cover against any such compensation claims.

Employers' liability insurance will enable you to meet the cost of compensation for your employees' injuries or illness, whether they are caused on or off site.

Injuries or illness relating to motor accidents which occur while your employees are working for you are dealt with under motor vehicle insurance and not Employers' Liability Insurance.

The law requires that you must be insured for a minimum of £5 million. This means that the level of compensation which could be paid out by the insurance company on your behalf could be up to £5 million *in any one year*. Employers' Liability Insurance is an *annual* requirement. However, many insurance companies will now automatically cover employers for up to £10 million in claims because the 'compensation culture' has driven awards to much higher levels.

It is perfectly legal to have more than one current employers' liability insurance cover, but each must be for a minimum claim level of £5 million.

The Employers' Liability (Compulsory Insurance) Regulations 1998 brought in the upper cover limits.

Do I need to tell my employees that I have Employers' Liability Insurance?

Yes. It is a legal requirement to display at the place of work the Employers' Liability Insurance Certificate. This states who the insurer is, when cover was taken out and when it expires and the value or limit of the overall cover in monetary terms.

The Certificate must be easy to find and read because it is the employees' right to choose to make a compensation claim if they feel you have contributed to their accident, injury or ill-health. They will need to know who to contact and be able do so directly without advising their employer first.

A claim under your Employers' Liability Insurance cover does not necessarily mean that it will be successful. The claim will be investigated by the insurance company or broker and they will advise you whether they believe settlement of the claim is justified or not. Any non-settlement may mean that the aggrieved person proceeds to the civil courts for a hearing.

Employers' Liability Insurance taken out in England is usually valid in Scotland, Wales, Northern Ireland, Isle of Man, Jersey and Guernsey.

Can my insurer make me pay, as employer, part of the compensation claim?

The insurance company must pay the whole amount of compensation agreed between the employee and themselves, or awarded by the court.

The insurance company cannot impose any conditions which mean that you have to pay towards the compensation. What the insurance company will obviously do if they have to pay claims out on your behalf is to raise the insurance premium. Remember, Employers' Liability Insurance is legally compulsory and so without adequate insurance cover you will not be able to trade as a business or profession if you need to employ people.

Any employer can agree with the insurance company to pay towards the insurance claim — this is known as the 'policy excess' and can vary from a few hundred to a few thousand pounds. The higher the excess you are prepared to take (usually) the lower the premium for the insurance.

Are any employers exempt from the legal requirement to have Employers' Liability Insurance?

Yes. The following employers are exempt:

- most public organisations including government agencies, departments, local authorities, police authorities
- health service bodies, e.g. NH Trusts, Primary Care Trusts, etc.
- some other organisations financed through public funds, e.g. passenger transport committees
- family businesses which only employ immediate members of the family, e.g. spouse, parents, children, grandchildren, brothers, sisters, etc.
- but any family business incorporated as a limited company must have Employers' Liability Insurance.

Does Employers' Liability Insurance need to cover anyone else other than employees?

Yes. People who are self-employed or freelance and who work for you irrespective of whether you pay them as self-employed people, may be 'employees' under the terms of insurance.

What matters, in all aspects of health and safety law when determining who is and who isn't an employee, is the relationship between you as employer and the individuals who work for you, and the degree of control you have over the work they do.

Generally, Employers' Liability Insurance is required for someone who works for you if:

- you deduct national insurance and income tax from the money you pay them
- you have the right to control where and when they work
- you supply most materials and equipment
- you have the right to any profit your workers may make even though you may share this with them
- you bear any losses they should make
- you require only that person to deliver the service — they cannot provide substitutes or sub-contract
- they are treated the same way as employees with regard to working conditions, etc.

Do volunteers, students on placements, etc. have to be covered by the insurance?

Not necessarily, but it would be good practice and advisable to do so.

Volunteers and students on placement who do not get paid for the work they do would not be 'employees' but they may be included under the insurance cover because you have control over the work that they do.

Is it essential to keep previous Employers' Liability Certificates?

Certificates of Insurance which have expired must be kept for 40 years after the expiry date. This is because many claims for industrial diseases are made many years after the disease is caused, e.g. asbestosis can take 10–20 years to become life-threatening and so knowing the insurance details of the employer 20 years ago will be vital to a claim for compensation.

What happens if I do not have Employers' Liability Insurance?

The law is enforced by the Health and Safety Executive and local authorities and when they make an inspection they will ask to see the Certificate of Insurance. It must show a minimum cover of £5 million.

Should an employer not have a valid Certificate of Insurance displayed and should they be unable to prove that they have adequate, current and valid cover then the law is contravened and an employer can be prosecuted.

Employers can be fined up to £2500 for each and every day they trade or practice their business without suitable cover.

Failure to display the Certificate or refusal to make it available to the Inspector carries a fine of £1000 maximum.

If the employer is a company, who really is responsible for health and safety?

A company is a legal entity and can be classed as equal to a 'person'. Should the company as an employer commit health and safety offences then the company is charged with those offences.

A company has to have specified positions with nominated people to fill those positions under company law. These specified individuals or positions are considered to represent the company in all matters.

All limited liability companies must have:

- at least one director, and
- a Company Secretary.

In any criminal investigation for breaches of health and safety law, the 'officers' of the company would represent the company.

So, the Managing Director, Chief Executive, Safety Director or Company Secretary would be expected to appear in court to act as the 'company'.

However, corporate responsibility is sometimes not a sufficient deterrent for unscrupulous employers and they have been able to abdicate any responsibilities for unacceptable health and safety conditions.

The Health and Safety Executive have considered introducing regulations which would require directors of companies to be personally liable for health and safety. While regulations have yet to be decided upon, the Health and Safety Executive did issue a 'Guidance Note' on Directors' Responsibilities for health and safety. This clearly sets out what is expected of company directors in undertaking their responsibilities for health and safety within their company.

What are the duties of Directors in respect of health and safety?

Directors collectively represent the company and must ensure that the company operates within the parameters of all laws.

Under Section 37 of the Health and Safety at Work Etc. Act 1974, a Director, Manager, Company Secretary or similar managerial position may be prosecuted *personally* if they consented to the commission of an offence, connived with the offence, i.e. turned a blind eye, or acted with neglect in respect of the offence.

The corporate body could be prosecuted for the offence and, also, an individual if they were found to have been personally responsible through consent, connivance or neglect.

The Guidance document on Director's responsibilities lists five Action Points which Directors must ensure are followed if they are to discharge their responsibilities. The five points are given below.

Action Point 1

The Board needs to accept formally and publicly its collective role in providing health and safety leadership in its organisation.

Action Point 2

Each member of the Board needs to accept their individual role in providing health and safety leadership for their organisation.

Action Point 3

The Board needs to ensure that all Board decisions reflect its health and safety intentions, as articulated in the health and safety policy statement.

Action Point 4

The Board needs to recognise its role in engaging the active participation of workers in improving health and safety.

Action Point 5

The Board needs to ensure that it is kept informed of, and remains alert to, relevant health and safety risk management issues. The Health and Safety Commission recommends that the Board appoint one of their number to be the 'health and safety director'.

What do the five Action Points mean in reality?

Companies and directors must take their responsibilities seriously.

Action Point 1

Everyone needs to know that, as a company and board of directors, you are serious about health and safety and that continuous improvement in health and safety performance is a key business indicator in your organisation.

You will need to explain your expectations and how your organisation and procedures will achieve them.

It is recommended that the public acceptance of health and safety responsibilities is included in Company Annual Reports.

Action Point 2

As a Board member, you will need to ensure that your actions and decisions taken at work reflect the intent of the Board's health and safety policy statement. Individuals in positions of responsibility must lead by example and health and safety must be seen as a key priority, both for the business and for the individual concerned.

Employees' belief in the commitment of an organisation to health and safety standards can be undermined by inconsistencies in actions by managers and directors.

Action Point 3

Many business decisions will have health and safety implications. Investment in new plant, equipment, premises and processes may fundamentally alter existing health and safety procedures and standards. Any decisions to embark on company expansion and changes must consider the ramifications for health and safety.

A high profile area of health and safety management is 'supply chain management'. Many organisations are seeking reassurances from their suppliers that they meet health and safety standards because it can be damaging to an organisation which prides itself on its own health and safety standards if it buys products or services from suppliers who may operate ‘sweat shop’ standards and illegal practices, etc.

Action Point 4

It is important to get your workforce on board with respect to health and safety standards, intentions, objectives, etc. There already exists a statutory duty to consult with employees on matters of health and safety but it is good practice to go even further and do more than the legal minimum.

Often, employees know what is best for them in respect of health and safety practices, type of personal equipment, etc. Your commitment to involving employees in health and safety may extend to extensive training initiatives, health campaigns, work–life balance campaigns, etc.

Action Point 5

The Board’s health and safety responsibilities must be properly discharged and the Board needs to put in place procedures for ensuring that this happens.

The Board will therefore need to:

- review the health and safety performance of the company regularly, i.e. at least annually
- ensure that the health and safety policy reflects the Board’s current priorities
- ensure that the management systems provide for effective monitoring and reporting of health and safety performance
- be kept informed about any significant health and safety failures, and the outcome of investigations into their failures
- ensure that it addresses health and safety considerations in all its decisions

- ensure that health and safety risk management systems are in place and remain effective. Periodic audits can provide information on their operation and effectiveness.

The Board Director who takes on the 'Health and Safety Director' title will be the person responsible for ensuring that health and safety is considered within the organisation and that systems and procedures remain relevant and effective. The 'Health and Safety Director' will not necessarily be the person immediately considered as being the person to prosecute in any contraventions of the law.

The Chief Executive/Chairman or Managing Director have a vital role to play in ensuring that the company provides adequate resources for the Health and Safety Director to take on the responsibilities.

What happens if no Board Director will take on the 'health and safety director' responsibilities?

The Chief Executive or Chairman of the Board will be expected to assume the overall responsibilities for health and safety. The 'Health and Safety' Director is not the only Board Director with these responsibilities — all Directors carry collective responsibility. The designated Director is merely to ensure that the company takes its duties seriously and puts in place a risk management programme to effectively monitor, review and improve health and safety standards and conditions.

Are there any other legal changes proposed to alter and improve Company Directors' responsibilities for health and safety?

March 2003 saw a '10 Minute Rule' Bill presented to Parliament on the subject of greater legal duties to be placed on company directors in respect of health and safety.

Case study

The Public Enquiry Report into the *Herald of Free Enterprise* tragedy stated that:

> The Board of Directors did not appreciate their responsibilities for the safe management of their ships. They did not apply their minds to the question. 'What orders should be given for the safety of our ships?' The Directors did not have any proper comprehension of what their safety duties were. There appears to have been a lack of thought about the way in which the *Herald* ought to have been organised for the Dover–Zeebrugge run.'

The report also criticised other levels of management within the shipping company but it concluded that only action by the company directors could have ensured that the company operated in a safe manner. Even if every employee in the company had, on the day of the disaster, done everything they should have done, the company would still have been operating a dangerous system which could only have been corrected by action on the part of the directors.

The Bill, tabled as The Company Directors' (Health and Safety) Bill imposes general health and safety responsibilities on directors. The Bill also requires companies to designate a particular director to be responsible for health and safety. Under the proposed law, they would have particular responsibilities such as monitoring health and safety and ensuring the right systems are in place within the company so as to enable such monitoring to take place. They would also be required to report significant health and safety failings to other directors, and they would be required to make recommendations for change.

The Bill has major Trade Union backing and is progressing through the parliamentary process during 2003.

What information might be relevant for inclusion in the Company's Annual Report?

You should include appropriate health and safety information in your published reports on your activities and performance in respect of health and safety. This demonstrates to your stakeholders your commitment to effective health and safety risk management principles.

Good corporate governance requires all companies to review their risk management policies, including health and safety. The Turnbull Report summarised the approach to be taken on this subject.

As a minimum, the company's Annual Report should contain the following in respect of health and safety information:

- the broad context of your policy on health and safety, including significant risks faced by your staff and the arrangements for consulting employees and involving safety representatives
- the company's health and safety goals, including specific and measurable targets for improving health and safety within your organisation.

Some specific and measurable targets may include:

- reducing working days lost due to injury or ill-health by *x*% per 100 000 worker hours
- reducing the incidence rate of work related ill-health by *x*% per number of employees
- reducing the major injury or fatality rate by *x*% by the year 2010
- a report on health and safety performance
- the number of reportable accidents, ill-health and dangerous occurrences notified to the authorities under the Reporting of Injuries, Diseases and Dangerous Occurrences Regulations 1995
- details of any major accidents, fatalities and the improvements implemented as a result of the investigation
- the total number staff days lost due to accidents or ill-health caused by work activities
- the number of health and safety enforcement notices served on the company within the reporting period
- the number and nature of convictions for health and safety offences
- the continuous improvement initiatives undertaken by the organisation
- the outcome of health and safety audits, internal and external monitoring processes, etc.
- the investment undertaken in respect of training and information initiatives for employees with regard to health and safety subjects, etc.

Increasingly, stakeholders in a business, including investors, want to have a clear understanding of the integrity, ethics and values of the companies they are dealing with. The new approach of 'Corporate Social Responsibility' expects companies be transparent in respect of their performance in all areas of social cultural and ethical dealings.

As an employer, am I responsible for the actions of my employees if they cause an accident to happen which results in injury to either themselves or others?

Yes, employers are responsible if persons are injured by the wrongful acts of their employees, if such acts are committed in the course of their employment.

This is know as 'vicarious liability'.

There is no vicarious liability if the act was not committed in the course of employment, e.g. one employee assaulting another is not something the employer would be liable for.

What is the 'duty of care'?

Every member of society is under a duty of care for something or someone.

Duty of care really means to take reasonable care to avoid acts or omissions which they can reasonably foresee are likely to injure their neighbour or anyone else who ought reasonably to be kept in mind.

Employers owe a duty of care to their employees.

Employers also owe a duty of care to contractors, visitors, members of the public, people on neighbouring properties, etc.

The duty of care owed by employers to employees includes:

- safe premises
- safe systems of work
- safe plant, equipment and tools
- safe fellow workers.

Other legislation also implies a duty of care to people, namely:

- The Occupiers' Liability Acts 1957 and 1984
- Consumer Protection Act 1987.

What is a Safety Policy?

Under the Health and Safety at Work Etc. Act 1974, employers must produce a written Health and Safety Policy if they have five or more employees.

The policy must contain a written statement of their general policy on health and safety, the organisation of the policy and the arrangements for carrying it out.

Employees must be made aware of the Safety Policy and must be given information, instruction and training in its content, use, their responsibilities, etc.

A copy of the Safety Policy Statement must either be given to all employees or be displayed in a prominent position in the workplace.

The Safety Policy must be reviewed regularly by the employer and kept up to date to reflect changes in practices, procedures, the law, etc.

What is meant by organisational arrangements?

This section of the Safety Policy shows how the organisation will put its good intentions into practice and outlines the responsibilities for health and safety for different levels of management within the company.

An organisational section will normally include:

- health and safety objectives
- responsibilities for:
    - Managing Director/CEO
    - Operations Director
    - Safety Director
    - senior management
    - departmental heads
    - maintenance
    - employees

- training arrangements
- monitoring and review processes
- appointment of competent persons
- consultation process for health and safety
- appointment of employee representatives
- procedures for conducting Risk Assessments
- emergency plans.

What is meant by 'arrangements' in a Safety Policy?

The Safety Policy must either contain details of what employees and others must do in order to ensure their safety at work, or it must contain references as to where information on safe practices can be found, e.g. in the department handbook, employee induction pack, etc.

Usually, however, for ease of use and clarity, most employers will produce everything needed for the Safety Policy in one document.

The 'arrangements' section of the Safety Policy contains the details of *how* you expect your employees and others, e.g. contractors, to proceed with a task or job activity safely.

Subjects often covered under arrangements are:

- accident and incident reporting and investigations
- first aid
- Risk Assessments
- Fire Risk Assessments
- manual handling
- using equipment
- electricity and gas safety
- personal protective equipment
- emergency procedures
- fire safety procedures
- training
- monitoring and review procedures

- COSHH procedures
- occupational health
- maintenance and repair
- permit to work procedures
- stress in the workplace
- violence in the workplace
- operational procedures.

A Safety Policy needs to be 'suitable and sufficient', not perfect.

If an accident happens in the workplace, the Investigating Officer (either HSE or EHO) will almost always want to see a copy of the Safety Policy and Risk Assessments. They will be looking to see if you had considered the hazard and risks of the job and implemented control measures. They will want to establish whether employees knew what to do safely and the best place to review such information is in the Safety Policy.

Prosecutions have been brought for failing to have a written Safety Policy and also for having a totally inadequate one.

The Safety Policy should be thought of as a communication tool between you, the employer, and your workforce. It should be their reference guide on how you expect them to perform their job tasks safely.

What are the five steps to successful health and safety management?

The five steps are:

Step 1: Set your policy
Step 2: Organise your staff
Step 3: Plan and set standards
Step 4: Measure your performance
Step 5: Audit and review

Step 1: Policy

- Do you have a clear Health and Safety Policy?
- Is it written down?
- Is it up to date?
- Does it specify who is responsible for what and who has overall safety responsibility?
- Does it give responsibilities to Directors and is there a clear commitment that the health and safety culture starts at the top?
- Does it specify arrangements for carrying out Risk Assessments, identifying hazards, implementation of control measures?
- Does it name competent persons?
- Does it state how health and safety matters will be communicated throughout the organisation?
- Does it have safety objectives?
- Has it had a beneficial effect on the business?
- Does it imply a pro-active safety culture within the organisation?

Step 2: Organise your staff

- Have you involved your staff in your health and safety policy?
- Are you 'walking the talk'?
- Is there a health and safety culture?
- Have you adopted the 'four Cs'?
    - competence
    - control
    - co-operation
    - communication.
- Are your staff and others competent to do their jobs safely?
- Are they properly trained and informed?
- Are there competent people around to help and guide them?
- Have you designated key people responsible for safety in each area of the business?

- Do employees know how they will be supervised in respect of health and safety?
- Do employees know who to report faults and hazards to and what will be done?

Step 3: Plan and set standards

- Have you set objectives with your employees?
- Have you reviewed accident records to see what general standards of health and safety you have?
- Have you set targets and benchmarks?
- Is there a purchasing policy regarding safety standards for equipment, etc.?
- Are there procedures for approving contractors?
- Have safe systems of work been identified?
- Have Risk Assessments been completed?
- Is the hierarchy of Risk Control followed?
- Have hazards to persons other than employees been assessed?
- Has a training plan and policy been developed and are there minimum levels of training for all employees?
- Are targets and objectives measurable, achievable and realistic?
- Is there an emergency plan in place?
- Have fire safety procedures been completed?
- Is there a 'zero tolerance' policy on accidents?
- Has the safety consultation process with employees been established?
- Is there a culture of continuous improvement in respect of health and safety?

Step 4: Measure performance

- Have you measured or can you measure with regard to your health and safety performance?

  - where you are
  - where you want to be
  - where the difference is
  - why?
- Are you practicing active monitoring or do you simply react to the situation when things go wrong?
- Is there a culture of recording near misses or do you wait for accidents to happen?
- Can you benchmark how you perform against another department, company or other organisation?
- Do you know how well you are really doing or does your performance just look good on paper?
- Are there ongoing accident and incident records and trend analysis?
- Is the effectiveness of the training measured — do you assess learning outcomes?
- Is there good legal compliance with health and safety law? Are you up to date with legal changes?

Step 5: Audit and review

- Are you regularly checking that the business is safe and minimising risks to health and safety?
- Is there a formal audit review process?
- Are Risk Assessments reviewed pro-actively?
- Are accidents investigated and processes changed as a result?
- Is there a formal audit process?
- Is it independent?
- Do staff get involved?
- Do you share the findings of the reviews?
- Is the Board kept up to date?
- Is there a health and safety committee?
- Is your business genuinely improving in respect of managing health and safety?

Managing health and safety is no different to managing any aspect of a business and it should be considered just as important as finance, sales, etc.

The likelihood of new corporate manslaughter legislation should focus the minds of all employers to address a thorough and comprehensive health and safety management system.

2

Legal framework

What is the main piece of legislation which sets the framework for health and safety at work?

The Health and Safety at Work Etc. Act 1974 is the main piece of legislation which sets out the broad principles of health and safety responsibilities for employers, the self-employed, employees and other persons.

The Health and Safety at Work Etc. Act 1974 is known as an 'enabling act' as it allows subsidiary regulations to be made under its general enabling powers. It sets out 'goal objectives' and was one of the first pieces of legislation to introduce an element of self-regulation.

The Act place responsibilities on employers to:

- safeguard the health, safety and welfare of employees
- provide a safe place of work
- provide safe equipment
- provide safe means of egress and access
- provide training for employees
- provide information and instruction
- provide a written Safety Policy
- provide safe systems of work.

The Act places responsibilities on employees to:

- co-operate with their employer in respect of health and safety matters
- wear protective equipment or clothing if required
- safeguard their own and others' health and safety
- not recklessly or intentionally interfere with or misuse anything provided in the interests of health and safety or welfare
- not tamper with safety equipment provided by their employers for the safety of themselves or others.

The Act places responsibilities on 'persons in control of premises' to:

- ensure safe means of access and egress
- ensure that persons using premises who are not their employees are reasonably protected in respect of health and safety.

The Act also requires employers to conduct their undertaking in such a way that persons who are not their employees are not adversely affected by it in respect of health and safety.

Finally, the Act requires manufacturers and suppliers and others who design, impart, supply, erect or install any article, plant, machinery, equipment or appliance for use at work, or who manufacture, supply or import a substance for use at work, to ensure that health and safety matters are considered in respect of their product or substance.

The term 'reasonably practicable' is used in the Health and Safety at Work Etc. Act 1974 and numerous Regulations. What does it mean?

The term 'reasonably practicable' is not defined in any of the legislation in which it occurs. Only the courts can make an authoritative judgement on what is reasonably practicable.

Case law has been built up over the years and practical experience gained in interpreting the law. A common understanding of 'reasonably practicable' is:

> the risk to be weighed against the costs necessary to avert it, including time and trouble as well as financial costs.

If, compared with the costs involved of removing or reducing it, the risk is small (i.e. consequences are minor or infrequent) then the precautions need not be taken.

Any establishment of what is reasonably practicable should be taken before any incident occurs.

The burden of proof in respect of what is reasonably practicable in the circumstances rests with the employer or other duty holder. They would generally need to prove why something is *not* reasonably practicable at a particular point in time.

An ability to *meet* the costs involved in mitigating the hazard and risk is *not* a factor which the employer can take into consideration when determining 'reasonably practicable'. Costs can only be considered in relation to whether it is reasonable to spend the money given the risks identified.

It is only possible to determine 'reasonably practicable' when a full, comprehensive Risk Assessment has been completed.

How can action under civil law be actioned by employees in respect of health and safety at work?

Civil action can be initiated by an employee who has suffered injury or damage to health caused by their work.

The employer may be in breach of his 'duty of care' which he owes to his employee. He may have been negligent in common law — that body of law which has been determined by case law — which has evolved rather than being set down by parliament.

Civil action may be brought on the grounds that the employer is in breach of statutory duty, i.e. he has failed to follow the requirements of statutory law (e.g. Acts and Regulations). Many Acts and Regulations do not necessarily confer a right to action in civil law if statutory duty is breached but some do, e.g. Construction (Design and Management) Regulations 1994.

How quickly do employees need to bring claims to the courts in respect of civil claims?

Civil actions must commence within *three* years from the time of knowledge of the cause of action. This will usually be the date on which the employee knew or should have known that there was a significant injury and that it was caused by the employer's negligence.

Therefore, an employer would be wise to keep all records of training, Risk Assessments, checks, maintenance schedules, etc. for a minimum of three years as these may be needed for any defence to a claim.

A civil claim will succeed if the plaintiff — the person bringing the case — can prove breach of statutory duty or the duty of care beyond 'the balance of probabilities'.

The employer may mount a number of defences to the claim, the most common of which are:

- contributory negligent — i.e. the injured employee was careless or reckless (e.g. ignored clear safety rules and procedures)
- injuries not reasonably foreseeable — i.e. the injuries were beyond normal expectation or control — the employer did not have the knowledge to foresee the risks; neither did science or experts
- voluntary assumption of risk — i.e. the employee consents to take risks as part of the job. But the employer cannot rely on this defence to excuse him of fulfilling his duties under legislation — no-one can contract out of their statutory duties.

What are the consequences for the employer if an employee wins a civil action?

Employers will invariably be required to pay damages or compensation. Compensation claims can run into thousands of pounds in some cases. The employers' Liability Insurance will cover the cost of the claims, less any excess which the employer opts to pay.

Damages are assessed on:

- loss of earnings
- damage to any clothing, property or personal effects
- pain and suffering
- future loss of earnings
- disfigurement
- inability to lead an expected, normal personal or social life because of the injury, etc.
- medical and nursing expenses.

There will not only be the financial payout but associated bad publicity which could lead to loss of reputation.

Who enforces health and safety legislation?

There are, in the main, two organisations with powers to enforce health and safety legislation:

- the Health and Safety Executive
- the local authorities.

The Health and Safety Executive enforce the law in the following types of work environment:

- industrial premises
- factories, manufacturing plant

- construction sites
- hospitals and nursing/medical homes
- local authority premises
- mines and quarries
- railways
- oil rigs
- broadcasting and filming
- agricultural activities
- shipping
- airports
- universities, colleges and schools.

The local authorities, usually through their Environmental Health Departments, are responsible for:

- retail sale of goods
- warehousing of goods
- exhibitions
- office activities
- catering services
- caravan and camping sites
- consumer services provided in a shop
- baths, saunas and body treatments
- zoos and animal sanctuaries
- churches and religious buildings
- childcare businesses
- residential care.

The powers both enforcement agencies have are the same but the way they use them may differ. Local authority inspectors may visit premises more frequently than the HSE but the HSE may take a tougher line on formal action because they do not have the resources to return to check for improvements.

What powers do enforcement authorities have to enforce health and safety legislation?

Inspectors can take action when they encounter a contravention of health and safety legislation and when they discover a situation where there is imminent risk of serious personal injury.

Inspectors can also instigate legal proceedings although in reality, the decision to proceed to court is often taken by the enforcement agency's in-house legal team.

Inspectors may serve an *Improvement Notice* if they are of the opinion that a person:

- is contravening one or more of the relevant statutory provisions

or

- has contravened one or more of those provisions in circumstances where the contravention is likely to occur again or continue.

The inspector must be able to identify that one or more legal requirements under an Act or Regulations are being contravened, e.g. failure to complete a Risk Assessment, operating an unsafe system of work.

An Improvement Notice must:

- state what 'statutory provisions' are, or have, been contravened
- state the manner in which the legislation has been contravened
- specify the steps which must be taken to remedy the provision
- specify a time within which the person is required to remedy the contraventions.

The time allowed to remedy contraventions highlighted in an Improvement Notice must be at least 21 days. This is because there is an appeal procedure against the service of an Improvement Notice and the appeal must be brought within 21 days.

An Improvement Notice served on a company as an employer must be served on the registered office and is usually served on the Company Secretary.

If an inspector believes that health and safety issues are so appallingly managed in workplaces or premises, etc. that there is 'a risk of serious personal injury' then a Prohibition Notice can be served. The inspector must be able to show or prove a risk to health and safety.

Prohibition Notices can be served in *anticipation* of danger and the inspector does not have to identify specific health and safety legislation which is being, or has been, contravened.

A Prohibition Notice must:

- state the inspector's opinion that there is a risk of serious personal injury
- specify the matters which create the risk
- state whether statutory provisions are being, or have been or will be, contravened and, if so, which ones
- state that the activities described in the Notice cannot be carried on by the person on whom the Notice is served, unless the provisions listed in the Notice have been remedied.

A Prohibition Notice takes effect immediately where stated, or can be 'deferred' to a specified time.

Risks to health and safety do not need to be imminent but, usually, there must be a hazard which is likely to cause imminent risk of injury, otherwise the inspector could serve an Improvement Notice.

What are the consequences for failing to comply with any of the Notices served by inspectors?

Failure to comply with either an Improvement Notice or a Prohibition Notice is an offence under the Health and Safety at Work Etc. Act 1974.

Legal proceedings are issued against the person, employer or company on whom the Notice was served and the matter will be heard in the magistrates' court in the first instance. However, serious contraventions can be passed to the Crown Court where powers of remedy are greater.

Failure to comply with an Improvement Notice carries a fine of up to £20 000 in the magistrates' court or an unlimited fine in Crown Court, plus possible imprisonment.

Failure to comply with a Prohibition Notice carries a fine of up to £20 000 in the magistrates' court, or an unlimited fine in the Crown Court. Imprisonment of persons who contravene a Prohibition Notice is also an available option for the courts — both magistrates' and Crown Courts.

Where employers or others fail to comply with a Prohibition Notice and continue, for example, to use defective machinery, and where, as a consequence of using that defective machinery, a serious accident occurs, there will invariably be a prosecution and the courts are likely to take a serious view and hand down prison sentences.

Can an inspector serve a Statutory Notice and prosecute at the same time?

Yes. If an inspector serves an Improvement Notice or Prohibition Notice they may decide that the contravention is so blatant or so serious that immediate prosecution is warranted. The Notices ensure that unsafe conditions are remedied during the prosecution process as legal cases can take several months to come before the courts.

Employers and others do not have to be given time to remedy defects once they have been identified as many duties on employers are 'absolute', i.e. the employer 'must' do something.

Inspectors will often prosecute without initiating formal action where they believe they have given the employer ample opportunity to put right defects, e.g. they may have given advice during a routine inspection, issued informal letters, etc.

Often, when accidents are being investigated, contraventions of health and safety legislation will be viewed seriously and prosecutions will frequently be brought.

What is the appeal process against the service of Improvement and Prohibition Notices?

Any person on whom a notice is served can appeal against its service on the grounds that:

- the inspector wrongly interpreted the law
- the inspector exceeded their powers
- the proposed solution to remedy the default is not practicable
- the breach of law is so insignificant that the notice should be withdrawn.

An appeal must be lodged within 21 days of the service of the Notice.

An appeal is to an Employment Tribunal.

An Improvement Notice is suspended pending the appeal process. This means that the employer or person served with the Notice will be able to continue to do what they have been doing without changing practices or procedures.

Costs for bringing an appeal may be awarded by the Tribunal — either in favour of the enforcement authority if the appeal is dismissed or for the employer if their appeal is successful.

A Prohibition Notice may continue in force pending an appeal unless the employer or person in receipt of the Notice requests the Employment Tribunal to suspend the Notice pending the appeal.

All Statutory Notices served by enforcement authorities should contain information on how to lodge an appeal.

On hearing the appeal, an Employment Tribunal can:

- dismiss the appeal, upholding the Notices as served
- withdraw the Notices, thereby upholding the appeal

- vary the Notices in respect of the time given to complete works or the remedies listed in any schedule
- impose new remedies not contained in the Notice if these will provide the solutions necessary to comply with the law.

If remedial works cannot be completed in the time given in the Improvement Notice, can it be extended?

Yes. It is allowable for the Inspector who served the Notice to extend the time limits specified if works cannot be completed and a request is submitted to the enforcing authority.

It would be prudent to explain why compliance cannot be achieved, what steps have been taken in the interim and when compliance can be expected.

What are the powers of inspectors under the Health and Safety at Work Etc. Act 1974?

Enforcing authorities and their individual inspectors have wide ranging powers under the Act and in addition to the service of notices can:

- enter and search premises
- seize or impound articles, substances or equipment
- instruct that premises and anything in them should remain undisturbed for as long as necessary while they conduct their investigation
- take measurements, photographs and recordings
- detain items for testing or analysis
- interview persons
- take samples of anything for analysis, including air samples

- require to see and copy if necessary, any documents, records, etc. relevant to their investigation or inspection
- require that facilities are made available to them while carrying out their investigation.

If inspectors believe that they will meet resistance, they may be accompanied by a police officer. It is an offence to obstruct an inspector while they are carrying out their duties.

What is a Section 20 interview under the Health and Safety at Work Etc. Act 1974?

Under Section 20, an inspector can require anyone to answer questions as they think fit in relation to any actual or potential breach of the legislation, accident or incident investigation, etc.

A Section 20 interview is, or should be, a series of questions asked by the inspector — not a witness statement where the interviewee describes what happens.

The answers to Section 20 questions must be recorded and the interviewee must sign a declaration that they are true.

However, evidence given in a Section 20 interview is inadmissible in any proceedings subsequently taken against the person giving the interview or statement, or his or her spouse.

If an inspector is contemplating bringing criminal proceedings, he will usually opt to interview a person under the Police and Criminal Evidence Act 1984 (PACE) as the information gathered during these interviews is admissible in court. The Code of Practice on PACE interviews is strict, e.g. a caution must be given and a failure to follow the procedures could result in acquittal on grounds of technicality.

What is the law on corporate manslaughter?

There are two types of manslaughter:

- voluntary — involving an intent to kill or do serious injury with mitigating circumstances, including provocation and diminished responsibility
- involuntary — for all other killings, usually sub-divided into two categories:
  - manslaughter by an unlawful or dangerous act
  - manslaughter by reckless or possibly gross negligence.

Companies can be prosecuted for manslaughter but it is difficult to prove because it is not always possible to establish that the 'directing mind' of the organisation, i.e. its directors and senior managers, were sufficiently aware of any safety contravention to be reckless, etc. Proving manslaughter charges is easier with small companies where, for instance, the Managing Director makes all the decisions and knows whether or not the company is complying with the law.

The Law Commission issued a report in 1996 which recommended a new central offence of 'corporate killing' committed where a company's conduct in causing the death falls far below what could reasonably be expected. It would not expect that the risk of death be obvious or that the company be capable of appreciating the risk. It would be sufficient to prove that the death had been caused by a company's failure in the way that its activities were managed and organised.

The Government issued a White Paper in 2000 which indicated that the recommendations of the Law Commission would be implemented into UK law. To date (2003) no new legislation has been enacted and the law on corporate manslaughter stands as currently interpreted, although, in May 2003, the Home Office confirmed its commitment to introducing legislation.

However, the Health and Safety Executive have issued improved guidelines on the responsibilities of directors and the courts are more active in seeing individual directors and managers prosecuted for health and safety offences.

What are the fines for offences against health and safety legislation?

Health and safety offences are usually 'triable either way' which means that they can be heard in the magistrates' courts or Crown Courts. There are some relatively minor offences which can only be heard in the magistrates' courts and some serious offences which can only be heard in the Crown Courts.

The sentencing powers of the two courts are different, with the Crown Court operating with a jury. Higher fines and imprisonment can be imposed by the Crown Courts.

Offences only triable in the magistrates' court include:

- contravening an investigation ordered by the Health and Safety Commission
- failing to answer questions under Section 20 of the Health and Safety at Work Etc. Act 1974
- obstructing an inspector in their duties
- preventing another person from answering questions or co-operating with an inspector
- impersonating an inspector.

Offences 'triable either way' include:

- failure to comply with any or all of Sections 2–7 of the Health and Safety at Work Etc. Act 1974
- contravening Section 8 of the 1974 Act – intentionally or recklessly interfering with anything provided for safety
- levying payment on employees or others for safety equipment, etc. contrary to Section 9
- contravening any of the Health and Safety Regulations made under the 1974 Act or other enabling legislation
- contravening the powers of inspectors in relation to the seizure of articles, etc.
- contravening the provisions and requirements of Improvement and Prohibition Notices

- making false declarations, keeping false records, etc. in relation to health and safety matters.

If cases are to be brought before the magistrates' courts (summary offences) they must be brought within six months from the date the complaint is laid, i.e. lodged at the magistrates' court and summonses issued.

The magistrates' courts have received guidance from the Court of Appeal that they must not hear cases where there have been serious breaches of health and safety law, fatalities or major injuries because their sentencing powers are not adequate. They should refer these for trial to the Crown Courts.

Fines can be imposed as follows:

- breaches of Sections 2–6 of the Health and Safety at Work Etc. Act:
  - magistrates' court — a maximum of £20 000 fine for each offence
  - Crown Court — an unlimited fine for each offence
- breaches of Improvement and Prohibition Notices:
  - magistrates' court — a maximum fine of £20 000 or imprisonment for up to six months, or both
  - Crown Court — unlimited fine, or imprisonment for up to two years, or both
- breaches of health and safety regulations and other sections of the 1974 Act:
  - magistrates' court — fines of up to £5000 per offence
  - Crown Court — unlimited fines per offence.

There are proposals to raise the level of fines and a private member's bill has been tabled in 2003 for increased fines. The proposal has not yet been actioned.

The HSE issues 'guidance' and Codes of Practice. What are these and is a criminal offence committed if they are not followed?

The HSE endeavours to provide employers and others with as much information as possible on how to comply with legislation.

Guidance documents are issued on a variety of specific industries or particular processes with the purpose of:

- interpreting the law, i.e. helping people to understand what the law says
- assisting in complying with the law
- giving technical advice.

Following guidance is not compulsory on employers, and they are free to take other actions to eliminate or reduce hazards and risks.

However, if an employer does follow the guidance as laid down in the HSE documents, they will generally be doing enough to comply with the law.

Approved Codes of Practice are the other common documents issued by the HSE and these set out good practice and give advice on how to comply with the law. A Code of Practice will usually illustrate the steps which need to be taken to be able to show that 'suitable and sufficient' steps have been taken in respect of managing health and safety risks.

Approved Codes of Practice have a special legal status. If employers are prosecuted for a breach of health and safety law and it is proved that they have *not* followed the relevant provisions of the Approved Code of Practice (known as an ACOP), a court can find them at fault unless they can show that they have complied with the law in some other way.

If an employer can show that they followed the provisions of an Approved Code of Practice it is unlikely that they will be prosecuted for an offence. Equally, if the employer follows the ACOP and the enforcing authority serve an Improvement or Prohibition Notice, the employer would have grounds to appeal the Notice.

Case study

Health and safety fines

In its third annual report *Health and Safety Offences and Penalties 2001/02*, the HSE report fines per industry sector as:

General fines levied across all sectors:	£10 million
Individual fines:	£8284
	(increased from £6226)
Construction — fines per offence:	£7564
Manufacturing — fines per offence:	£9083
Extractive industries — fines per offence:	£17 550
Service sector — fines per offence:	£8832
Highest fines for health and safety offences imposed on corporate bodies:	£750 000
	£350 000
	£250 000
	£225 000

The HSE prosecuted 55 individuals, including cases against 31 directors.

3

Accident and incident management

What are the requirements of the Reporting of Injuries, Diseases and Dangerous Occurrence Regulations 1995?

Where any person dies or suffers any of the injuries or conditions specified in Appendix 1 of the Regulations, or where there is a 'dangerous occurrence' as specified in Appendix 2 of the Regulations, as a result of work activities, the 'responsible person' must *notify the relevant enforcing authority*.

Notification must be by telephone or fax and confirmed in writing within ten days.

Where any person suffers an injury not specified in the Appendix but which results in an absence from work of more than three calendar days the 'responsible person' must notify the enforcing authority in writing.

The 'responsible person' may be the employer, the self-employed, someone in control of the premises where work is carried out or someone who provided training for employment.

Where the death of any person results within one year of any notifiable work accident, the employer must inform the relevant enforcing authority.

When reporting injuries, diseases (e.g. industrial diseases contracted as a result of work undertaken such as Weil's Disease, miner's lung) or dangerous occurrences, the approved forms must be used — either F2508 or F2508a.

Records of all injuries, diseases and occurrences which require reporting must be kept for at least three years from the date they were made.

Accidents to members of the public which result in them being taken to hospital as a result of the work activity must be reported.

Incidents of violence to employees which result in injury or absence from work must be reported.

Why must these types of accident be reported?

National accident statistics are collated by the Health and Safety Executive in order to indicate the general state of health and safety across the country. Fatalities, major injuries and 'over three day' injuries are all recorded and allocated to industry-specific sectors so that the state of legal compliance, accident trends, etc. can be judged.

However, the most important reasons for notifying accidents are:

- it is a legal requirement
- so that the enforcing authorities can investigate to establish whether the employer has contravened the law
- so that serious incidents can be prevented from happening again.

Accident statistics for 2000/2001 are:

- 249 fatalities
- 27 477 major injuries
- 127 084 'over three day' injuries

- 384 fatalities to members of the public (including suicides on railways)
- 14 362 non-fatal injuries to members of the public.

If an accident is reported to the enforcing authority, will an investigation take place?

Not always. It depends on the severity of the accident and the approach of the enforcing authorities.

Any accident which involves a major injury is highly likely to be investigated as it shows to the authority that something serious may have gone wrong with the employer's safety management system.

Sometimes, an enforcing authority will make a telephone investigation first and request further details of management systems so that they can assess your general attitude and commitment to health and safety. If they find information inadequate they will make a site visit.

What are the consequences if I ignore the law on reporting accidents?

A failure to notify accidents, diseases and dangerous occurrences is an offence under the Regulations and the number of prosecutions for non-compliance is rising. Fines are up to £5000 in the magistrates' court.

What are the types of accident which have to be notified?

It is not actually accidents which have to be notified, but the consequences of those accidents and the type of injuries which they cause.

Accidents and incidents which arise out of or in connection with work and which fall into the category of:

- fatality
- major injury
- 'over three day' injury

must be reported.

Also, any accident or incident which involves a member of the public or non-employee being sent to hospital also needs to be reported. This is so that information can be gathered on how safe work practices are for members of the public using premises, etc.

Certain types of 'dangerous occurrence' must also be reported. These would be incidents which have the potential to cause major or multiple injuries and which could affect large numbers of people (i.e. high risk catastrophes).

Industrial diseases must also be reported within twelve months of the disease being identified.

What are major injuries?

A major injury must be reported to the enforcing authority. Any accident at work or caused by the work activity which results in the following is notifiable:

- any fracture of a bone (other than finger, thumb or toes)
- any amputation
- dislocation of the shoulder, hip, knee or spine
- loss of sight (whether temporary or permanent)
- a chemical or hot metal burn to the eye or any penetrating injury to the eye
- any injury resulting from an electric shock or electrical burn (including one caused by arcing) leading to unconsciousness or requiring resuscitation or admittance to hospital for more than 24 hours

- injuries leading to hypothermia, heat-induced illness or unconsciousness, or requiring resuscitation or requiring admittance to hospital for more than 24 hours
- loss of consciousness caused by asphyxia or by exposure to a harmful substance or biological agent
- acute illness requiring medical attention or loss of consciousness resulting from the absorption of any substance by inhalation, ingestion or through the skin
- acute illness requiring medical treatment where it may be caused by exposure to a biological agent, its toxins or infected material.

What are 'over three day' injuries?

Where a person at work is incapacitated for *more than* three consecutive days from their normal work owing to an injury resulting from an accident at work, then the accident must be reported.

The day the accident happens does *not* count in calculating the three days. But any days which would not be normal working days, e.g. shift days, days off, holiday, weekends, *do* count in the three days.

If an employee remains at work but cannot carry out their usual work, i.e. are put on 'light duties' then the accident *must* still be notified.

Does it matter when accidents are reported or are there strict time-scales?

As you would expect, there are strict time-scales for the reporting of accidents as follows:

- fatalities — immediately or as soon as possible after they happen

- major injuries — immediately or as soon as possible after they happen
- 'over three day' injuries — within *ten days* of them happening
- accidents to people who are not at work — immediately or as soon as possible after they happen
- dangerous occurrences — immediately or as soon as possible after they happen
- diseases — as soon as apparent and without undue delay.

Any accident, disease or dangerous occurrence which is notified immediately — usually by telephone, email or fax — must be confirmed in writing on the appropriate form within *ten days*.

Where and to whom should accidents be notified?

Overall, the number of notifications of accidents is generally low and there is serious under-reporting (hence the trend to prosecute for non-compliance).

In order to address this, the HSE have made the reporting of accidents much easier and have provided a 'one-stop shop' for all employers to report accidents, etc. irrespective of whether their enforcing authority is the HSE or the local authority.

All notifiable accidents, dangerous occurrences and diseases can be notified to the Incident Contact Centre (ICC) on:

Telephone:	0845 300 9923 (office hours only Monday–Friday)
Fax:	0845 300 9924
Internet:	www.riddor.gov.uk
Post:	Incident Contact Centre Caerphilly Business Park Caerphilly CF83 3GG

The ICC will forward the details of the accident to the appropriate authority.

The ICC will send out confirmation copies of any notifications made.

If reports are made on-line, a reference number will be given and the report will be printed out.

What forms have to be used for notification?

There is a standard form for use under RIDDOR, known as F2508.

Blank forms can be printed from the RIDDOR web site.

On-line notifications use the F2508 template.

When you phone the ICC they will ask questions and obtain information in accordance with the layout of the F2508.

However, not having the correct form is not an excuse for failing to notify an accident.

Do records of accidents, etc. need to be kept by the employer?

Yes. Records of injuries and dangerous occurrences must be kept by the 'responsible person' for at least three years.

Records must contain:

- date and time of the accident or dangerous occurrence
- if an accident is suffered by a person at work:
  - full name
  - occupation
  - nature of the injury
- if an accident is suffered by a person not at work:
  - full name
  - status, e.g. customer, etc.
  - nature of injury

- place where the accident or dangerous occurrence happened
- a brief description of the circumstances
- the date that the event was first reported to the enforcing authority
- the method by which the event was reported.

Keeping copies of the F2508 will suffice.

Many employers have an accident book but the information contained in a standard accident book will not satisfy the detailed level of information needed if the accident is notifiable.

During 2003, the type of accident book used by many businesses must change, because personal information on individuals is available to anyone who wants to read the accident book and this is contrary to the Data Protection Act.

Accident books should therefore have one page for each entry and must not display personal data to anyone not authorised to read it.

The RIDDOR Regulations refer to the responsible person as being responsible for reporting accidents. Who is this?

The responsible person under the Regulations is:

- the employer
- the self-employed
- those in control of work premises where work is carried out.

If self-employed persons are injured while at work in other people's premises, then the person in control of the premises, i.e. employer, managing agent, facilities management company, will need to notify the accident.

What diseases need to be notified under the Regulations?

The number and type of diseases which need to be notified are many and varied.

Generally, any disease which is caused by an activity or work, or from being exposed to substances in use at work, will be notifiable.

Common diseases include:

- mesotheiomia
- asbestosis
- industrial deafness
- carpal tunnel syndrome
- leptospirosis (Weil's disease)
- legionnaires' disease
- anthrax
- brucellosis
- hand/arm vibration syndrome (vibration white finger)
- silicosis
- bladder cancer
- pneumoconiosis
- skin cancer.

What are dangerous occurrences?

The list of dangerous occurrences is quite long and is included in a Schedule to the RIDDOR Regulations.

An indication of dangerous occurrences is given below:

- collapse, overturning or failure of load-bearing parts of lifts and lifting equipment
- accidental release of biological agent likely to cause severe human illness
- accidental release of a substance which may damage health

- explosion, collapse or bursting of a vessel or associated pipework
- an electrical short circuit or overload causing fire or explosion
- an explosion or fire causing suspension of normal work for over 24 hours
- the collapse of scaffolding.

Any dangerous occurrence which has the potential to cause significant harm must be checked to see whether it is notifiable.

It is always better to err on the side of caution and report than not.

Does any injury which happens to a customer or member of the public have to be reported?

No. Only those accidents and resultant injuries which require a member of the public to be taken to hospital as a result of the accident, and which were caused by the employer's *work activity*.

Injuries which happen to customers or others which are due to carelessness or from something over which they have control will not be notifiable.

The injury must result from an accident 'arising out of or in connection with work'.

Types of incident which would *not* be reportable if they caused injury to a person:

- acts of violence causing injury between fellow workers over a personal argument
- a customer dying of a heart attack on the premises
- a visitor tripping over their own bag or luggage
- acts of violence between customers or visitors.

Is there a legal duty to investigate accidents?

No, not at present although formal guidance is expected during 2003/04 on how to investigate accidents.

Accident reporting is only one part of the process of health and safety management. When an accident or incident occurs it is necessary to find out what caused it, what went wrong, why it went wrong and what can be done to ensure that it does not happen again.

The HSE consulted widely during 2001 on the subject of accident investigation as they felt that employers were not doing enough about accident prevention and management. The HSE were considering a new set of Accident Investigation Regulations, amendments to existing Regulations, e.g. RIDDOR and the Management of Health and Safety at Work Regulations 1999, an Approved Code of Practice or Guidance Notes.

The consultation process has ended with the HSE deciding to issue formal Guidance on Investigating Accidents and to deal with the matter informally as 'best practice' rather than making it a legal duty.

There is, in any event, implied requirements and duties to investigate accidents because Risk Assessments have to be reviewed regularly and when circumstances change. An accident may be 'changed circumstances'.

Also, with the increase in civil claims, insurance companies are forcing employers to investigate the causes of accidents so that strategies can be put in place to prevent future occurrences. This will help to bring down Employers' Liability Insurance premiums or, at the very least, prevent them rising astronomically.

What are the key steps in an accident investigation?

Every employer should have an accident investigation plan as part of their health and safety management policy.

Accident investigation should be looked upon as identifying what happened, and why, so that a reoccurrence can be prevented.

Step 1: define the purpose of the investigation
Step 2: define the procedure
Step 3: define what equipment will be needed

An Accident Investigation Kit

Contents:

- report form
- routine checklist for basic questions or prompts to the investigator
- notebook, pad, paper, pen
- tape recorder for on-site comments or to assist at interviews
- camera — instant or digital to take immediate photos of the scene of the accident
- measuring tape, e.g. surveyor's tape, builder's tape
- any special equipment regarding the work environment which could assist the investigation, e.g. noise meters, air sampling kits
- witness forms for statements.

Step 4: define how the investigation is to be carried out and what information will need to be gathered
Step 5: define the content of the report
Step 6: decide how recommendations will be implemented.

It is sensible to create an 'accident investigation kit' so that everything you need is in one place and valuable time is not lost in trying to find equipment.

What issues will I need to consider when an accident investigation is carried out?

If there is an accident investigation procedure in place you should follow the specific guidelines.

If not, some pointers aregiven below.

- Where did the accident happen — describe exactly?
- Who was injured — were they an employee, contractor, member of the public?
- What were they doing?
- What equipment were they using?
- What time was it?
- What were the environmental conditions?
- What was the condition of the area or equipment?
- Are any defects noted, e.g.:
  - maintenance issues
  - worn flooring or other trip hazards
  - poor lighting
  - broken guarding?
- Was there a safe system of work in place?
- Where Risk Assessments available?
- Had COSHH Assessments been completed?
- Who witnessed what happened?
- What did they see happen, or what did they hear?

- What training had the employee had?
- What actions were taken immediately after the incident?
- Was the accident notifiable?
- What needs to be done to prevent it happening again?
- Had equipment been routinely checked?
- Are maintenance records available?
- Had the work process been regularly reviewed and checked as part of safety monitoring?
- Did someone not do something they should have done?
- Were contractors involved? Had they had induction training and been made aware of any site-specific hazards?

4

First aid

What is the main piece of legislation which covers first aid at work?

The Health and Safety (First Aid) Regulations 1981 set the standards for first aid at work.

The main scope of the Regulations covers:

Every employer must provide equipment and facilities which are adequate and appropriate in the circumstances for administering first aid to employees.

Employers must make an assessment to determine the needs of their workplace. First aid precautions will depend on the type of work and, therefore, the risk being carried out.

Employers should consider the need for first aid rooms, employees working away from the premises, employees of more than one employer working together and non-employees.

Once an assessment is made, the employer can work out the number of first aid kits necessary by referring to the Approved Code of Practice.

Employers must ensure that adequate numbers of 'suitable persons' are provided to administer first aid. A 'suitable person' is someone trained in first aid to an appropriate standard.

In appropriate circumstances the employer can appoint an 'appointed person' instead of a first aider. This person will take charge of any situation, e.g. call an ambulance, and should be able to administer emergency first aid.

Employers must inform all employees of their first aid arrangements and identify trained personnel.

What is first aid at work?

First aid at work covers the arrangements you must make to provide employees with adequate first aid attention while they are at work.

Employees may suffer injury or ill-health while at work, caused by a work activity or the work environment, or employees may become ill for other reasons while at work.

They must receive immediate emergency attention and, in serious cases, an ambulance must be called.

First aid at work is designed to save lives and to prevent minor injuries or incidents escalating into serious ones.

As an employer, what do I need to do?

The Health and Safety (First Aid) Regulations 1981 require employers to provide adequate and appropriate equipment, facilities and personnel to enable first aid to be given to employees if they are injured or become ill at work.

The Regulations set out some minimum first aid provisions to be provided on an employer's site as follows:

- a suitably stocked first aid kit
- an appointed person to take charge of any incident and the first aid arrangements
- first aid facilities to be available at all times when people are at work.

The key words regarding first aid are:

- 'adequate and appropriate'
- 'suitable and sufficient'.

Satisfaction of the two requirements will be dependent on the actual circumstances of the workplace.

In order to ascertain what appropriate first aid is required, employers will have to carry out a Risk Assessment of first aid needs.

What should an employer consider when assessing first aid needs?

The Risk Assessment process requires employers to consider the hazards and associated risks involved in work activities.

Small businesses will need only the simplest of Risk Assessments and basic first aid provision.

Larger businesses will need to consider the following.

Step 1

Consider what are the risks of injury and ill-health associated with your work practices and activities.

Step 2

Are there any specific risks which can be clearly identified, e.g.:

- working with hazardous chemicals
- working with dangerous tools
- working with dangerous machinery
- working with dangerous loads
- working with animals.

Step 3

Are there areas within the business where risks may be greater because of the environment and might these need additional first aid facilities, e.g.:

- research laboratories
- pathology laboratories
- hot working environments
- cold working environments.

Step 4

Consider the businesses record of accidents and injuries. What types have occurred? How serious have they been? Is there any evidence to show that extra precautions, etc. are necessary? Where and when did they happen, and why?

Step 5

How many people are employed on the site? How many are permanent, temporary, etc.? How familiar are they with procedures, processes, etc.?

Are there any young or inexperienced workers more likely to have an accident or suffer ill-health than the regular workforce. Does anything extra need to be done for people with disabilities.

Step 6

What are the buildings used for and how are they used, are they spread out, do employees work out of doors, can they access all parts of the building? Where might first aid provisions and facilities be located so that they are available to all.

Step 7

Do employees work outside of normal working hours, work overtime, work alone, etc.? How might they raise the alarm? Do employees travel frequently?

Step 8

Would emergency services be easy to summon? Could they gain access to the site or building if it were out of normal hours? Are work places accessible?

What are some of the steps an employer needs to take in order to ensure suitable first aid provision?

The assessment of the likelihood and frequency of injury and ill-health will determine what will need to be done to ensure suitable first aid provision, i.e. that emergency aid is provided.

First aid kits will need to be provided — should there be one in a central location or will it be preferable to have several smaller ones easily accessible to employees?

An appointed person or first aider will need to be appointed depending on the number of employees and severity of risk of injury, etc.

First aid room facilities may be needed depending on the number of employees.

Emergency procedures to call the emergency services will be needed.

Special consideration will be needed for people with disabilities.

The type of first aid equipment will need to be considered, e.g. will eye-wash stations be needed?

Other workers and employees will need to be considered — who is providing first aid facilities for them, e.g. on construction sites?

What is an 'appointed person'?

An appointed person is someone who is not necessarily trained in emergency first aid treatment but who is appointed to take charge of an incident and call the emergency services, etc.

The appointed person usually also looks after the first aid equipment and ensures that it is adequately stocked, in the right location, etc.

Employers who employ up to 50 employees and whose business falls into the category of 'low risk', i.e. office environment, libraries, retail shops, etc. need only appoint an 'appointed person'.

There should also be 'reserve' appointed persons to cover for holidays and sickness.

What is a first aider?

A first aider is someone who has undergone training in first aid and holds a current first aid at work certificate. All first aider training courses must be approved by the HSE, so ask all providers for details of their HSE registration documentation, etc.

Employers who have over 50 employees usually need to appoint a trained first aider.

First aid training courses last four days and the certificate is valid for three years. After that, retraining is needed.

A trained first aider can administer first aid and the primary purpose is to prevent injuries from getting worse rather than to try to treat people or provide medical expertise.

How many first aiders or appointed persons does an employer need?

The Regulations do not set down hard and fast rules in respect of numbers of people to be appointed. It really is dependent on the type of work activity and the likelihood of injury.

The law requires the employer to make the assessment and, as long as that can be explained and justified as being suitable and sufficient, the law will be satisfied.

However, the Approved Code of Practice on First Aid at Work gives guidance on suitable numbers of appointed persons and first aiders.

Enough people should be nominated and trained so that absences can be covered.

A suggested ratio of appointed persons or first aiders is listed as follows.

Shops, offices, libraries	Less than 50 employees	1 appointed person
Shops, offices, libraries	50–100 employees	1 first aider
Shops, offices, libraries	Over 100 employees	1 first aider, plus an extra person per 100 employees
Food processing, warehouses	Fewer than 20 employees	1 appointed person
Food processing, warehouses	20–100 employees	1 first aider for every 50 employees
Food processing, warehouses	More than 100 employees	2 first aiders plus an extra person for every 100 employees
Construction sites, industrial sites, manufacturing, spray shops, chemical industries, etc.	Less than 5 employees	1 appointed person
Construction sites, industrial sites, manufacturing, spray shops, chemical industries, etc.	5–50 employees	At least 1 first aider
Construction sites, industrial sites, manufacturing, spray shops, chemical industries, etc.	More than 50 employees	A first aider for every 50 employees

What should be in a first aid kit?

There is no legal list of items which should be in a first aid box, although there is guidance in the First Aid Approved Code of Practice.

The contents of a first aid kit really depend on the Risk Assessment for first aid facilities which every employer must complete.

First aid kits should *not* however, contain any medicines, e.g. aspirin or paracetamol, because no one is trained to issue medicines and the recipient could be allergic to the substance, etc.

The contents of a first aid kit are relatively straightforward and the following would be sensible contents for a low risk work environment:

- 20 individual sterile adhesive dressings (plasters) of varying sizes
- 2 sterile eye pads
- 4 sterile triangular bandages
- safety pins
- 6 medium-sized sterile wound dressings
- 2 large-sized sterile wound dressings
- disposable gloves
- an advice leaflet.

All dressings, plasters, etc. should be individually wrapped. Dressings have a 'shelf life' and dates should be checked and anything out of date should be replaced because it may no longer be sterile.

Is an employer responsible for providing first aid facilities for members of the public, customers etc.?

No, the law on first aid applies to employees while they are at work.

However, it is good practice to consider the needs of customers or the public when completing the Risk Assessment. The Health and

Safety Executive strongly recommend that public and customers are included in first aid provision.

Are there any other responsibilities which an employer has in respect of first aid?

An employer must inform their employees of their first aid arrangements. This is a legal requirement.

Notices can be displayed advising where the first aid kit is located, who the appointed persons or first aiders are.

Special arrangements will need to be considered for any employees with language problems, learning disabilities, etc.

When is it necessary to provide a first aid room?

There is no legal requirement within the Health and Safety (First Aid) Regulations 1981 for employers to provide a first aid room. As with the provision of any first aid facilities, the need for a first aid room will be determined in the Risk Assessment.

Guidance in the Approved Code of Practice indicates that it would be good practice to provide a first aid room when there are 150 employees or more.

If a first aid room is provided, it must have the following attributes:

- easily accessible to all employees
- easily accessible for emergency services and for stretchers, etc.
- provided with heating, lighting and ventilation
- provided with hand-washing facilities and, preferably, a sink
- provided with drinking water
- provided with a chair, couch, table/desk, etc.
- provided with first aid materials — first aid kits, etc.
- provided with a refuse bin

- contain blankets and pillows, etc.
- have some method of raising the alarm
- contain record books
- surfaces which are easily cleaned and disinfected
- clean and tidy
- provided with suitable first aid information, e.g. names of first aiders, etc.

A first aid room should be exclusively used as a first aid room so that it is available in any emergency but if it needs to be used as a workplace, procedures should be in place to ensure that its use as a first aid room will not be prejudiced.

What records in respect of first aid treatment, etc. need to be kept by the employer?

It is considered good practice to keep records of all incidents which require any first aid treatment or attendance by a first aider.

A record book should be available to record the following:

- date, time and place of the incident
- injured person's name and job title
- a description of the injured person's injuries or illness
- details of first aid treatment given
- details of any actions taken after treatment given, e.g. did employee or person go to hospital, go home, etc.
- name and signature of person who gave first aid treatment or who oversaw the incident.

Consideration must be given to any data protection requirements and to ensure the anonymity of personal information. It would be a sensible idea to have a new page for each person treated and for previous entries to be kept in a secure drawer, etc. People should not be able to 'flip through' the first aid record book and learn personal facts about colleagues or others.

Case study

A young employee arrived at work in a call centre and, after about an hour or so, she reported that she did not feel well, had a headache and felt sick. The supervisor called the first aider who suggested that she would probably benefit from a 'lay down' in the first aid room. She was taken to the first aid room and settled down on the couch, still feeling quite poorly.

Her colleagues were quite busy and forgot all about her being in the rest room until about lunch time. A colleague went to visit the first aid room and found that she was in a coma. An ambulance was called and the young woman was taken to hospital. Unfortunately, she died later that day from a brain haemorrhage.

Could anything have been done differently?

Yes, the first aider who attended the young woman should have been responsible for her care while she was in the first aid room and should have visited her every 30 minutes or so. If she had not felt better after, say, an hour, she should have been taken to her GP or to the casualty department at hospital. Records of the checks carried out on her should have been kept. She might still have died, but she should not have been left for several hours unattended.

5

Manual handling

What are employers responsible for with respect to manual handling at work?

The Manual Handling Operations Regulations 1992 apply to all manual handling activities carried out by employees while at work.

Employers must, as far as is reasonably practicable, avoid the need for employees to undertake any manual handling activities while at work which involve risk of injury.

Despite the above requirement, manual handling is a common cause of work-related injury. In some cases, poor manual handling can lead to permanent disability and physical impairment.

Employers must undertake a Risk Assessment of all manual handling activities and determine a hierarchy of risk control in order to minimise injury and ill-health risks.

Information of the weight of objects to be manually handled must be given to employees. This can be general information or more specific and precise product information. Many manufacturers and suppliers of products and equipment are displaying the weight of the item on packaging or on delivery notes, etc.

What are the costs of poor manual handling to both businesses and society?

Injuries sustained by employees while they are handling, lifting or carrying items at work account for 38% of all notified 'over three day' injuries.

Over 1 million people are reported to have suffered 'illness' from musculo-skeletal disorders and the prevalence rate is increasingly when compared with the early 1990s.

Statistics from the HSE for 2001/2002 show that approximately 12.3 million days were lost in employment productivity due to musculo-skeletal disorders.

The National Health Service has one of the highest incidence rates for musculo-skeletal injury and approximately 50% of all reported 'over three day' injuries were due to injury while handling, lifting and carrying, i.e. approximately 5000.

Back injury is not the only type of injury to be sustained from manual handling. Injuries are reported which affect:

- hands
- feet
- arms
- legs
- neck
- head.

Many manual handling injuries are the result of poor practices being followed over lengthy periods of time and not from a 'one-off' manual handling activity.

On average, each injury takes 20 days for recovery and, in some instances, disability is permanent. Costs to business will be huge and are often hidden in real terms. Costs to be considered are:

- costs of sick pay
- cost of loss of skilled employee
- replacement/temporary staff
- reduced productivity

Case study

An ambulance worker received compensation in 2002 of £140 000 in an out-of-court settlement with his employers for serious back injuries sustained in the course of his employment.

The employee was lifting a patient when two wheels came off the stretcher he was carrying. He then had to bear the patient's weight for five minutes.

As a result, he damaged his lower back and right leg.

Damages were claimed against the NHS Trust because the stretcher had been modified to fit into the ambulance and was not fit for its purpose. The Trust admitted liability.

Consideration is being given by the NHS Trust to instigating legal action against the stretcher manufacturer.

- investigation time
- civil claims
- criminal prosecution
- increased insurance premiums.

Back injuries represent the biggest single group of claims for incapacity benefit.

Costs for manual handling injuries have been estimated at £6 billion in lost production.

What steps should be taken in respect of carrying out a Risk Assessment for manual handling?

In the first instance, it would be sensible to conduct a general assessment to see if manual handling activities give rise to hazard and risk, as not *all* manual handling will.

Remember that manual handling includes:

- lifting
- pushing
- pulling
- shoving
- lowering
- carrying.

Consider the size and shape of the load and the best way to handle it:

- If the load is difficult or heavy, seek assistance.
- Consider where the load is going — is the pathway clear and free from obstruction.
- Is the place ready where the load is to go?
- Can lifting devices be used?
- Can the load be split to make carrying easier?
- Manual handling involves pushing and pulling as well as lifting. Can any of these jobs be mechanised?
- Complete a manual handling Risk Assessment.

What needs to be considered for a detailed Manual Handling Risk Assessment?

The tasks

Do the tasks involve:

- holding loads away from the body
- twisting, reaching or stooping
- strenuous pushing or pulling
- unpredictable movement of loads
- large vertical movement
- long carrying distances

- repetitive handling
- insufficient rest time
- a work rate imposed by a process?

The loads

Are the loads:

- heavy, bulky or unwieldy
- difficult to hold
- unstable or unpredictable
- intrinsically harmful, e.g. sharp?

The working environment

In the working environment, are there:

- constraints on posture
- variations in level
- poor floors
- hot, cold or humid conditions
- strong air movements
- poor lighting
- restrictions on movement or posture from clothes or personal protective equipment?

Individual capacity

Does the job:

- require unusual capability
- endanger those with a health problem
- endanger pregnant women
- require special information or training?

Case study

Types of manual handling in licensed premises

The following activities undertaken routinely in most pubs are likely to present particular risks from manual handling operations:

- the delivery and removal of full and empty kegs, boxes, barrels, crates and gas cylinders
- the stacking of full kegs and barrels
- the movement of kegs, barrels, etc. within the cellar or storeroom
- shifting of casks
- the movement of loads between floors — carrying crates from the cellar to the bar
- lifting buckets of water or pipe cleaning containers
- lifting gas cylinders
- putting items on shelves and getting items off shelves
- movement of furniture, equipment, etc.
- food deliveries
- removal of glass bottle skips
- carrying empty glass baskets or crates
- changing optics
- lifting glass washer trays
- carrying tills or money drawers
- carrying money or change
- moving AWP machines
- assisting with entertainment equipment.

What are some of the ways of reducing the risks of injury from manual handling?

Some ways of reducing the risk of injury are detailed below.

The tasks

Can you:

- reduce the amount of twisting and stooping
- avoid lifting from floor level or above shoulder height
- avoid strenuous pushing or pulling
- reduce carrying distances
- avoid repetitive handling
- vary work, allowing one set of muscles to rest while another is used?

The loads

Can you make the loads:

- lighter or less bulky
- easier to hold
- more stable
- less damaging to hold
- have you asked your suppliers to help?

The working environment

Can you:

- improve workplace layout to improve efficiency
- remove obstructions to free movement
- provide better flooring

- avoid steps and steep ramps
- prevent extremes of hot and cold
- improve lighting
- consider less restrictive clothing or personal protective equipment?

Individual capacity

Can you:

- take better care of those with physical weaknesses or who are pregnant
- give your employees more information, e.g. about the range of tasks they are likely to face
- provide training?

What are good handling techniques?

The following are important points to bear in mind when handling a load, using a basic lifting operation as an example.

Planning

Plan the lift. Where is the load to be placed? Use appropriate handling aids if possible. Do you need help with the load? Remove obstructions. For a long lift, such as floor to shoulder height, consider resting the load midway on a table or bench to change grip.

Positioning feet

Feet should be apart, giving a balanced and stable base for lifting (tight skirts and unsuitable footwear make this difficult). Leading leg should be as far forward as is comfortable and, if possible, should be pointing in the direction you intend to go.

Good posture

When lifting from a low level, bend with the knees, but do not kneel or over-flex the knees. Keep the back straight, maintaining its natural curve (tucking in the chin helps). Lean forward a little over the load if necessary to get a good grip. Keep the shoulders level and facing in the same direction as the hips.

Holding the load

Try to keep the arms within the boundary formed by the legs. The best position and type of grip depends on the circumstances and individual preference, but must be secure. A hook grip is less tiring than keeping the fingers straight. If you need to vary the grip as the lift proceeds, do it as smoothly as possible.

Lifting

Keep the load close to the trunk for as long as possible. Keep the heaviest side of the load next to the trunk. If a close approach to the load is not possible, slide it towards you before you try to lift. Lift smoothly, raising the chin as the lift begins, keeping control of the load.

Movement

Move the feet instead of twisting the trunk when turning to the side.

Adjustment

If precise positioning of the load is necessary, put it down first, then slide it into the desired position.

What are the guideline weights for lifting or manual handling?

		Female:	Male:
(1)	Shoulder height		
	— arms extended	3 kg	5 kg
	— near to body	7 kg	10 kg
(2)	Elbow height		
	— arms extended	7 kg	10 kg
	— near to body	13 kg	20 kg
(3)	Thigh height		
	— away from body	10 kg	15 kg
	— near to body	16 kg	25 kg
(4)	Knee height		
	— away from body	7 kg	10 kg
	— near to body	13 kg	20 kg
(5)	Lower leg height		
	— away from body	3 kg	5 kg
	— near to body	7 kg	10 kg

Each category indicates the guideline weights for lifting and lowering loads.

Heavier weights can be handled more safely if they are held close to the body. Carrying objects at arm's length creates extra strain of the spine and muscles and therefore lower weights are recommended.

The weights assume that the load is readily grasped with both hands and that the operation takes place in reasonable working conditions with the lifter in a stable body position.

Any operation involving more than twice the guideline weights should be rigorously assessed — even for fit, well-trained individuals working under favourable conditions.

Twisting

Reduce the guideline weights if the lifter twists to the side during the operation. As a rough guide, reduce them by 10% if the handler twists beyond 45°, and by 20% if the handler twists beyond 90°.

Frequent lifting and lowering

The guideline weights are for infrequent operations — up to about 30 operations per hour — where the pace of work is not forced, adequate pauses to rest or use different muscles are possible, and the load is not supported for any length of time. Reduce the weights if the operation is repeated more often. As a rough guide, reduce the weights by 30% if the operation is repeated five to eight times a minute; and by 80% where the operation is repeated more than twelve times a minute.

Exceeding the guidelines

The guidelines are not necessarily safe limits for lifting. But exceeding the guidelines is likely to increase the risk of injury, so you should examine the work closely for possible improvements. You should remember that you must make the work less demanding if it is reasonably practicable to do so.

Is 'ergonomics' anything to do with manual handling?

Yes and no! Ergonomics is the science concerned with the 'fit' between people and their work and surroundings.

Ergonomics aims to make sure that tasks, equipment, information and the environment suit each worker. So, that could include manual handling activities but it is more likely to consider *how* the job is done rather than *what* is lifted.

How can ergonomics improve health and safety?

Applying ergonomic principles to the workplace will:

- reduce the potential for accidents
- reduce the potential for injury and ill-health
- improve performance and productivity.

Equipment, controls, operating panels, isolation switches, etc. should all be designed for ease of use but, in practice, how many times are switches awkward to get at, requiring twisting and contortion to use them?

A machine with a control panel which the operator is required to use could be the cause of accidents and injury if:

- the switches and buttons could be easily knocked on or off, thereby starting or stopping the machine by mistake
- the warning lights or switches are unusual colours or the opposite colours to those usually expected, e.g. if red is for 'go', green is for 'danger'. Also, colours may be important as many people have colour blindness for red and green
- the instruction panel and information given on how to use the controls is complicated or too detailed, causing operator confusion and inappropriate actions.

Ergonomics would look at all of the above issues and 'design out' the hazards associated with the control panel on the machine. The location of controls would be considered so as to cut down on 'repetitive strain' injuries.

What kind of manual handling problems can ergonomics solve?

Ergonomics is typically thought of as solving physical problems and in respect of manual handling these problems would be:

- loads which are too heavy or bulky
- loads which need to be lifted from the floor or from above shoulder height

- repetitive lifting
- tasks which involve awkward postures, twisting or bending
- loads which cannot be gripped properly
- tasks which need to be carried out in poor environmental conditions, i.e. wet floors, poor lighting, cramped space, restricted headroom
- tasks carried out under too great time pressures and without adequate rest periods.

Any of the above situations can lead to operator tiredness and exhaustion. This increases the risk of accidents and injury.

Ergonomics is about finding solutions for alternative ways of doing the job.

6

Hazardous substances

What are the COSHH Regulations 2002?

The Control of Substances Hazardous to Health Regulations 2002 are known as COSHH.

The Regulations set out the duties that employers have to their employees and others to protect them from exposure to and harm from hazardous substances.

The 2002 Regulations came into force in November 2002 and replace all earlier sets of Regulations, i.e. 1988, 1994 and 1999.

The Regulations were amended in March 2003 to address further issues in respect of carcinogens.

What does COSHH require?

Five basic principles of occupational hygiene underline the COSHH Regulations:

(1) identify the hazardous substance, identify how it is to be used, assess the risk to health, precautions and health risks arising from that substance

(2) if the substance is harmful, wherever possible, substitute a less harmful substance
(3) introduce appropriate measures to prevent or control risks and ensure that control measures are used, that any protective equipment is properly maintained and that any safety procedures are observed
(4) where necessary, monitor the exposure of employees and introduce an appropriate form of surveillance of their health
(5) inform, instruct and train employees in the risks to their health and safety and the precautions that need to be taken.

What substances are covered by the COSHH Regulations?

The Regulations cover a wide range of substances and include those which are very toxic, harmful, corrosive, irritant or biological. These could include cleaning materials for floors, toilets, drains, glasswasher and dishwasher detergents, pest control materials, dusts, fumes, solvents, building products, oils, etc.

There is no limit to the application of COSHH with regard to the quantity of materials. The overriding principle is that if a substance is a hazard to health, it *must* be assessed.

All of these substances are safe when properly used, but the use of each must be assessed. Employees must be made aware of any hazards, the precautions necessary and trained how to use the substances correctly.

What changes did the 2002 COSHH Regulations introduce?

The 2002 Regulations did not fundamentally change employers' duties to ensure that employees and others are not exposed to the harmful effects of hazardous substances.

The Regulations generally have made changes as follows:

- to numerous definitions within the Regulations, e.g. biological agents, inhalable and respiratable dust
- COSHH Assessments under Regulation 6 have been amended to:
    - require that the steps identified by the Assessment as necessary to meet the requirements of the Regulations are implemented
    - the Assessment is to consider
        - the hazardous properties of the substance
        - information on health effects provided by the supplier, including information contained in the Safety Data Sheets
        - the level, type and duration of exposure
        - the circumstances of the work, including the amount of substance involved
        - activities, such as maintenance, where there is potential for a high level of exposure
        - any relevant occupational exposure limit or standard, maximum exposure limit or similar occupational exposure limit
        - the effect of preventative and control measures which have been or will be taken to comply with Regulation 7
        - the results of relevant health surveillance
        - the results of any monitoring of exposure
        - the risks of exposure to more than one substance, i.e. the 'cocktail' effect
        - the approved classification of any biological agent
        - such additional information as the employer may need to complete the Assessment
- the Assessment is to be reviewed if the results of monitoring show it to be necessary

- employers who employ five or more employees are to record the significant findings of the assessment as soon as is practicable after the Risk Assessment is made and steps are to be taken to implement control measures
- a specific requirement is introduced under Regulation 7 to substitute a substance or process if this eliminates or reduces risks to health
- control measures are listed in order of priority in Regulation 7
- biological agents are now covered in the body of the Regulations
- all control measures are to be kept clean
- new provisions regarding employee monitoring, the keeping of records and health surveillance have been introduced
- information, instruction and training requirements have been extended to include details on occupational exposure limits, access to relevant safety data sheets, exposure and health risks of the substance, significant findings of the COSHH Assessment, results of health surveillance, control measures to be implemented, etc.

Duties extend to training persons other than the employer's employees if those people are so exposed to the risks from hazardous substances.

How do dangerous chemical products get into the body?

There are three main ways in which products get into the body: through ingestion, through the skin or through inhalation. The form of the product plays an important role. The more finely divided a product is, the more easily it is absorbed (generally the smaller the particles, the more dangerous they are). Solids for example, may be in the form of powder and liquids in the form of an aerosol.

Absorption is dependent on many factors, including the state of subdivision of the product (i.e. the smallness of the particles), its concentration, the length of exposure, the use of protective equipment, its solubility in fat, etc.

Digestive route (via the mouth)

Entry via the digestive route (or ingestion) is usually accidental or the result of carelessness, for example:

- through transferring a product from one container to another by sucking it up through a pipette, or through a product having been stored in a food and drink container
- through eating, smoking, drinking, etc. after having handled a dangerous product and not having washed hands.

Percutaneous route (entry via the skin)

Certain products, such as irritant and corrosive products act locally at the place where they come into contact with the skin, the mucous membrane or the eyes.

Others, which are soluble in fat, not only act on the skin but also penetrate it and spread throughout the body where they can cause various disorders. This is the case with solvents, which degrease the skin, but which can also damage the liver, nervous system or kidneys. Benzene can damage the bone marrow. Motor fuel (which has a relatively high benzene content) should not be used to wash hands.

Small cuts and grazes provide an easy route for dangerous chemicals.

Respiratory route (entry via the lungs)

This is the most common entry route at work, as pollutants can be present in the atmosphere. They then enter the lungs with the air we breathe. This can occur when handling solvents, paints or glues, stripping leaded paint with a blow torch or welding, for example.

Once inhaled into the lungs, these chemicals enter the bloodstream and can cause damage not only to the respiratory system but also to the rest of the body.

A chemical which enters via any of these routes can be transported to other parts of the body in the bloodstream and can cause damage to other organs.

What are Occupational Exposure Limits or Standards (OEL/OES)?

For a number of commonly used hazardous substances, the Health and Safety Commission has assigned occupational exposure limits (or standards) to help define what is adequate control.

Occupational Exposure Limits are set at levels which will not damage the health of employees exposed to the substance by inhalation, day after day.

Where a substance has an OEL, the exposure of employees to the substance must legally be reduced to the OEL level.

What are maximum exposure limits?

Maximum exposure limits are set for substances which can cause the maximum amount of health damage. These substances usually cause life-threatening illnesses such as cancer, asthma, severe industrial dermatitis, respiratory conditions, etc.

Substances which have an MEL must be used only if there is no alternative and exposure must not exceed the stated limit over the given exposure time — usually no more than ten minutes.

Employers should avoid the use of all substances with an MEL — find an alternative.

What is health surveillance?

Health surveillance is required under certain circumstances and requires employers to assess the health of their employees regularly. If employees are exposed, for instance, to a substance which causes skin irritation, then it may be necessary to check the condition of hands and arms by visual examination from time to time.

Health surveillance allows an employer the opportunity to monitor the effectiveness of the control measures in place.

If employees are exposed to breathing in fumes or dust, then routine lung tests or blood tests can be used.

Health surveillance can be carried out by a medical doctor or occupational nurse, or an employer can carry out simple assessments and refer to experts for advice.

Hazardous symbols

See table overleaf.

Symbol	Meaning	Description of risks
	Toxic (T) Very toxic (T+)	Toxic and harmful substances and preparations posing a danger to health, even in small amounts. If very small amounts have an effect on health the product is identified by the toxic symbol.
	Harmful (Xn)	These products enter the organism through inhalation, ingestion or the skin.
	Highly flammable (F) Extremely flammable (F+)	(F) Highly flammable products ignite in the presence of a flame, a source of heat (e.g. a hot surface) or a spark. (F+) Extremely flammable products can readily be ignited by an energy source (flame, spark, etc.) even at temperatures below 0°C.
	Oxidising (O)	Combustion requires a combustible material, oxygen and a source of ignition; it is greatly accelerated in the presence of an oxidising product (a substance rich in oxygen).

	Corrosive (C)	Corrosive substances seriously damage living tissue and also attack other materials. The reaction may be due to the presence of water or humidity.
	Irritant (Xi)	Repeated contact with irritant products causes inflammation of the skin and mucous membranes, etc.
	Explosive (E)	An explosion is an extremely rapid combustion. It depends on the characteristics of the product, the temperature (source of heat), contact with other products (reaction), shocks or friction.
	Dangerous for the environment (<<N)	Substances which are highly toxic for aquatic organisms, toxic for fauna, dangerous for the ozone layer.

What are the recommended steps when undertaking a COSHH Assessment?

The HSE recommend an *eight step* approach to a COSHH Assessment as follows:

Step 1: assess the risks
Step 2: decide what precautions are needed
Step 3: prevent or adequately control exposure
Step 4: ensure that control measures are used and maintained
Step 5: monitor exposure of employees (and others if appropriate)
Step 6: carry out appropriate health surveillance
Step 7: prepare plans and procedures to deal with accidents, incidents and emergencies
Step 8: ensure that employees are properly trained, informed and supervised.

Who should carry out the COSHH Assessment?

As an employer, the responsibility for the assessment remains with you. However, the task of conducting and completing the Risk Assessment can be delegated but you cannot abdicate responsibility.

Whoever carries out the Risk Assessment will have to be able to demonstrate competency by:

- understanding the COSHH Regulations
- having access to the Regulations, Approved Codes of Practice and Guidance
- being able to obtain all the relevant information about the substances, e.g. Safety Data Sheets
- having knowledge of the effects of the substance on individuals and how to interpret findings

- having experience in the work processes and how substances are actually used in the workplace, e.g. this may be different from the manufacturer's/supplier's instructions due to workplace practicalities and custom, e.g. a product may be sprayed rather than painted
- having experience to be able to make decisions on risk.

Employees often know huge amounts about what they use, how and why. Consult with them. Use Employee Representatives or Safety Representatives for information and advice.

If you appoint external assistance, check out their experience and reputation. Take references.

What are some COSHH control procedures?

The COSHH Assessments have been devised to indicate the level of risk to health from use of and exposure to any of the hazardous chemicals in use within the company.

When a product is to be used, especially for the first time, consult the COSHH Assessments.

Follow the information given and make sure that staff are familiar with it.

Follow the control measures indicated, e.g. wear gloves, goggles, etc.

Do not forget that there could be people in the vicinity who might be affected.

Do not forget to plan for spillages.

Chemicals not in the COSHH Assessments should not be used.

Ensure that all staff know where the COSHH Assessments are kept for use in emergencies.

When the task is complete, return the chemical container, securely capped, to a suitable storage area.

How might the risk of accidents from hazardous substances be reduced?

- Check that packages and containers are in good condition, so as to avoid leaks. Make sure the gases, fumes, vapours or dusts are extracted at their point of origin. Wear a respirator if necessary. Watch out for possible sources of fire.
- Keep dangerous products only in appropriate containers, property labelled. Never transfer them into bottles such as lemonade or beer bottles, or other food containers. This type of practice causes serious accidents every year. Dangerous products should preferably be kept locked away when not in use.
- Avoid contact with the mouth. Do not eat, drink or smoke when using dangerous substances or when in a place where they are used.

7

Asbestos

As an employer, what are my key duties under the Control of Asbestos at Work Regulations 2002?

Employers have duties under the Regulations to protect their employees from exposure to asbestos-containing materials as they may cause harm to health.

Under the Regulations, employers are responsible for the health and safety of:

- their employees
- other peoples' employees
- members of the public
- the self-employed

if they are or will be exposed to asbestos.

Employers must also:

- provide information, instruction and training
- carry out Risk Assessments
- produce a written plan of work
- ensure that asbestos types are identified
- prevent or reduce exposure to asbestos
- introduce control measures

- maintain effective control measures
- keep records of any tests or examinations
- provide suitable protective clothing
- provide changing facilities and clean clothing
- develop emergency procedures
- prevent or reduce the spread of asbestos
- clean equipment and premises after exposure to asbestos
- designate areas as 'respirator zones' or an 'asbestos area'
- display suitable hazard warning notices
- arrange for effective air monitoring
- keep records for 5 years, or for 40 years if to do with health surveillance
- provide health surveillance every two years to those exposed to asbestos
- remove asbestos waste under special waste provision.

Contravention of any of the Regulations is an offence and fines can be up to £5000 per offence in the magistrates' court or, for serious offences and breaches of Improvement Notices or Prohibition Notices, unlimited with prison sentences in the Crown Court.

What is asbestos and what are the risks from it?

Asbestos is a natural mineral fibre which has been used for decades as an effective heat insulator, fire retardant material and general bonding material.

There are three main types of asbestos fibre:

- chrysotile = white
- amosite = brown
- crocidolite = blue.

Historically, blue and brown asbestos have been considered the most dangerous but recent research has indicated that all types of asbestos fibres are potential health hazards and so all types of asbestos have been banned for new building works and asbestos is no longer being imported or sold in the UK.

However, there is much residual asbestos in existence in buildings of all descriptions but, in particular, in those built up to about the mid 1980s. Brown and blue asbestos was banned in 1985 and white asbestos was banned in 1999.

Asbestos fibres cannot be easily visually identified by their colour and laboratory analysis is needed to positively identify the type and volume of fibres.

Asbestos fibres can become airborne and the microscopic fibres can be breathed into the lungs where they lodge for many years and, in so doing, create an irritant and cause lung damage or lead to even more serious diseases.

Breathing in asbestos fibres can lead to the individual developing one of three fatal diseases:

- asbestosis — scarring of the lung tissue leading to shortage of breath and difficulty in breathing
- lung cancer
- mesothelioma — cancer of the lining around the lungs and stomach.

There are no cures for asbestos-related diseases.

People who smoke and who are exposed to asbestos fibres are at an even greater risk of developing asbestos-related diseases.

Exposure to minute amounts of asbestos fibres over prolonged periods of time will increase the risks of disease.

Asbestos-related diseases take from 10 to 60 years to develop from first exposure and so it may be in retirement that the health effects of exposure are felt.

Are any employees or workers at greater risk than others?

Obviously, those who choose to work in the asbestos removal industry are exposed to a significant hazard, but the risk to which they are exposed may be quite minimal because the controls and checks placed upon the removal industry are considerable and, generally, workers are well protected.

Most of the individuals suffering from asbestos-related illnesses work in the building and maintenance trades.

Employees (or self-employed, etc.) who are most of risk are:

- plumbers
- electricians
- builders
- carpenters
- roofing contractors
- gas fitters and service engineers.

How does asbestos get into the body?

Asbestos fibres can be taken into the body via:

- breathing in fibres
- swallowing fibres.

Airborne fibres can be inhaled via both the mouth and nose. Larger asbestos fibres are generally filtered out by the lungs but the microscopic fibres are not.

Ingestion occurs when hands become contaminated with asbestos fibres and the individual wipes their mouth, eats food or otherwise transmits the fibres via the 'hand to mouth' route.

Is asbestos cement as much of a worry as asbestos fibres?

No, not quite such a worry but it is classified as a hazardous substance because the fibres which are held in place with the cement can become loose and airborne.

What is the new duty of care to manage asbestos?

The Control of Asbestos at Work Regulations (CAW) 2002 include a new duty for the management of asbestos-containing materials.

It is *not* illegal to have asbestos-containing materials in a place of work because if asbestos is in good condition and not releasing fibres into the atmosphere it is generally quite safe. In fact, there is less risk in leaving it in place than there is from removing it.

However, until the 2002 Regulations, there was no duty on anybody to maintain existing asbestos materials in good condition, nor was there any responsibility to monitor the condition so as to identify deterioration.

Regulation 4 of the CAW Regulations 2002 places legal responsibilities for managing asbestos on to duty holders.

A duty holder is defined in the Regulations as:

- every person who has, by virtue of a contract or tenancy, an obligation of any extent in relation to the maintenance or repair of non-domestic premises, or any means of access thereto or egress therefrom; or
- in relation to any part of non-domestic premises where there is no such contact or tenancy, every person who has, to any extent, control of that part of those non-domestic premises or any means of access thereto or egress therefrom.

Where there is more than one duty holder, the relative contribution to be made by each person in complying with the requirements of the Regulation will be determined by the nature and extent of the maintenance and repair obligation owed by that person.

A wide range of people will potentially have obligation under the Regulation, including:

- employers
- self-employed
- owner of premises
- managing agents.

The duty to manage asbestos does *not* extend to domestic premises but does apply to residential premises if they are let as a business, e.g. hotels, bed and breakfast establishments, caravan parks, etc.

As a duty holder what do I need to do?

Find out if asbestos-containing materials are present in your building or premises.

If the building was constructed before 1985 it is likely to contain some asbestos unless it has already been removed.

Buildings constructed up to 1999 may have asbestos cement materials.

How do I find out whether asbestos is present?

Checking building plans and other information such as operating manuals, etc. to see if any reference has been made to asbestos or asbestos-containing materials.

Consult the Design Team who undertook the building works, including any known contractors or sub-contractors, building services contractors, etc.

Carry out a full survey of the premises to identify likely asbestos-containing materials.

How do I carry out an asbestos survey?

With great caution! You may be able to carry out a visual inspection yourself in simple premises, but it is sensible to employ competent surveyors to do the job.

It helps to know where asbestos-containing materials *might* be so start with a survey in these high risk areas:

Ceiling voids	May have sprayed asbestos lagging or coating to structural timbers for fire protection, or loose asbestos material may be packed into voids as fire breaks.
Pipework	Sprayed asbestos coating or lagging is often found around heating pipes as both insulation and fire protection.
Boilers	Asbestos gaskets and seals are used for insulation and fire protection.
Electrical switchgear	Asbestos gaskets and seals are used for insulation and fire protection.
Ductwork, structural steels, firebreaks, etc.	Sprayed asbestos coating or lagging is used for insulation and fire protection.
Insulating boards	Asbestos cement insulation is contained in many boards for fire protection and insulation, e.g. on soffit boards, ceiling panels, partition walls, etc.
Ceiling boards	Lay-in grid ceiling tiles can contain asbestos.
Roofing felt	May contain asbestos.
Guttering and rainwater goods	May contain asbestos cement.
Roof coverings	May contain asbestos cement.
Water tanks	May be made of asbestos cement.
Floor tiles	May contain asbestos for both insulation and fire protection.
Fire doors	Often contain asbestos insulating board for fireproofing.

Assume that materials contain asbestos unless evidence is available to the contrary.

Do not disturb and break into material to see if it contains asbestos — you often cannot tell by looking at it and if it is asbestos you expose yourself and others to health risks from fibres.

If your initial survey indicates that materials are likely to contain asbestos then you will need to have a full destructive survey carried out whereby samples of the material are taken for analysis. Only trained and competent persons should undertake sampling.

What do I do if I find asbestos?

Assess its condition byapplying the following observational tests:

- Is the surface of the material damaged, frayed or scratched?
- Is any part of the material peeling or breaking off?
- Is it detached or loose from the structure it was applied to, e.g. falling away from pipes or structural steel?
- Are protective coatings and coverings damaged?
- Is asbestos debris or dust evident in the area?

What if it is in bad condition?

Asbestos in poor condition is a serious health hazard. You must either:

- remove it
- repair it
- encapsulate it
- seal it.

Licensed contractors will need to be appointed to work on asbestos. You must agree an Action Plan with them and ensure that all legal requirements are met.

What do I do after it has been encapsulated, sealed or repaired?

Keep records of what was done, where, by whom and when. An Asbestos Management Register should be created to help ensure that all asbestos is located and regularly inspected.

Display hazard warning signs on all residual asbestos or provide some other form of identification.

Operate Permit to Work procedures for maintenance works, refurbishment works, etc.

It is *not* illegal to have asbestos-containing materials on the premises, provided they do *not* cause a health hazard. It is safer to leave asbestos which is in good condition in situ and manage the potential risk, than remove it and create high health risks.

What do I do if asbestos is in good condition?

Leave it in place and label it accordingly. Hazard warning signs or colour coding will suffice. Make sure that people know what the signs or colours mean.

Keep an Asbestos Register which identifies the *location and condition* of the asbestos. Ensure that this is freely available to those who need the information.

An annotated plan of the premises would be ideal.

What do I have to do to manage the asbestos?

Make regular checks of the condition of the asbestos material to ensure that it is not deteriorating.

Carry out Risk Assessments for any work in the area, e.g. could anything puncture the asbestos material thereby releasing fibres, etc.?

Keep records up to date and show that an Asbestos Management Plan is in place.

Introduce a Permit to Work system for *all* contractors, maintenance engineers, etc. so that you know where they will be working, why, on what and what the hazards and risks are. If they are to work near asbestos material this can be highlighted and safety precautions stipulated.

How often do I need to check the condition of asbestos material?

Once every twelve months will usually be sufficient but if the asbestos is in an area which is at risk of damage, then more frequent inspections will be required.

Asbestos is in an area where I propose to carry out refurbishment works. What do I need to do?

In these instances, asbestos must be removed by a licensed contractor. There must be no risk of accidental asbestos fibre release so asbestos must be removed *before* works commence.

As a landlord, I have carried out my duties and produced an Asbestos Register. What responsibilities do my tenants have?

You must make sure that the information in the Asbestos Register is available to everyone who may need to know about this residual health hazard in the building.

You must ensure that any tenant is aware of their responsibilities as an 'employer' under the Control of Asbestos at Work Regulations

2002. They must not, for instance, expose any of their employees to risk and cannot therefore require works to be undertaken on asbestos. Maintenance and repair works must be managed.

If tenants, or employees, damage otherwise 'safe' asbestos, they may legally be responsible for the repairs.

Lease agreements should be reviewed so that they include clauses on asbestos management and specific responsibilities, etc.

How do I dispose of asbestos waste?

Asbestos comes under the Special Waste Regulations 1996 and can only be removed to a licensed waste disposal site.

Asbestos must be double bagged, sealed in heavy duty polythene bags and clearly labelled with a recognised asbestos label.

What are the Asbestos (Licensing) Regulations 1983?

These Regulations require that a contractor doing more than one hour's work per week with:

- asbestos insulation
- asbestos coating
- asbestos insulating board

or if the total work done by all workers is over two hours, must hold a licence issued by the HSE to remove asbestos.

Licensed contractors are subject to stringent inspections by the HSE's Asbestos Unit and they must meet expectations regarding:

- work plans
- method statements
- health surveillance

Checklist

Find	Check if materials containing asbestos are present.
Condition	Check what condition they are in.
Presume	Assume material contains asbestos unless there is strong evidence that it does not.
Identify	Arrange to have any asbestos-containing material analysed and identified by a specialist laboratory.
	Material must be identified if any refurbishment works, repair works, etc. are to be carried out in the area in which it is located.
	Material must be identified if it is in poor condition.
Record	Record the location and condition of asbestos containing material on a plan or drawing and describe its condition.
Assess	If the material is likely to deteriorate or to be disturbed it will be advisable to remove the material completely.
Plan	Prepare and implement a plan to manage the risks.

Actions

Good condition materials

↓

- Monitor at regular intervals.
- Label the material.
- Inform workers, contractors, service engineers, etc. of existence of material.

Minor damage to material

↓

- Repair and/or encapsulate.
- Monitor at regular intervals.
- Label the material.
- Inform workers, contractors, service engineers, etc. of existence of material

Poor condition of material

↓

- Asbestos in poor condition should be removed.

Asbestos disturbed

↓

- Asbestos likely to be distributed should be removed.

(Taken from HSE Information Leaflet INDG 223 rev 11/02.)

- welfare facilities
- training
- Risk Assessments
- disposal procedures.

Work with asbestos cement, asbestos-backed floor tiles, asbestos sarking felt, etc. does not need to be undertaken by a licensed contractor. However, there is still a risk from these materials and it would be best practice to engage the services of licensed contractors for all asbestos work.

Work with asbestos usually needs to be notified to the HSE (unless it is very minor) by way of the *fourteen day notice*.

Specific information has to be submitted to the HSE as follows:

- name of notifier
- address and telephone number of his usual place of business
- a brief description of:
    - the type of asbestos to be used or handled
    - the maximum quantity of asbestos to be held at any one time on the premises in which the work is to take place
    - the activities or process involved
    - commencement date of work activity.

What type of asbestos surveys are there?

There are three types of asbestos survey usually undertaken to identify asbestos in buildings. The type of survey chosen depends on circumstances but, before major refurbishment or demolition works are undertaken, it is essential to have the most comprehensive of surveys completed.

Type 1: Location and assessment or presumptive survey

The purpose of this survey is to locate, as far as is reasonably practicable, the presence and extent of asbestos-containing materials, including their condition.

Samples and analysis of materials is generally not undertaken until such time as more detailed information is required.

All areas should be accessed and inspected.

If it is unclear as to whether material contains asbestos it is to be assumed that it does until proved otherwise.

Type 2: Standard sampling, identification and assessment or sampling survey

Under this survey type, asbestos-containing materials are positively identified through sampling and analysis. Visual inspection is carried out as for Type 1, but samples from each type of asbestos-containing material are taken by the surveyor.

Results of the analysis will identify exactly what is and is not asbestos material. A management plan can then be actioned.

Type 3: Full access, sampling and identification or pre-demolition/major refurbishment survey

This survey is undertaken before all major building works as it requires that an extensive survey of the building is undertaken, with full destructive inspection if necessary, e.g. removal of ceiling panels or walls to identify any materials in loft voids, cavities, etc.

A full sampling programme is undertaken and asbestos materials are identified in all locations, together with their type and condition.

Volume and surface area of asbestos must be calculated as the information is often used on the Fourteen Day Notification Notice.

8

Noise

As an employer, why do I need to worry about noise?

Sounds and noise are an important part of everyday life. In moderation they are harmless but if they are too loud they can permanently damage your hearing. The danger depends on how loud the noise is and how long you are exposed to it. The damage builds up gradually and you may not notice changes from one day to another, but once the damage is done, there is no cure. The effects may include the following:

- sounds and speech may become muffled so that it is hard to tell similar sounding words apart, or to pick out a voice in a crowd
- permanent ringing in the ears (called tinnitus)
- a distorted sense of loudness. Sufferers may ask people to speak up then complaining that they are shouting
- needing to turn up the television too loud, or finding it hard to use the telephone.

The law requires employers to safeguard the health, safety and welfare of their employees while they are at work. The employer must provide a working environment which is safe and without risks to health.

So, any environment which is subject to excessive noise will be unsafe because there is a risk of noise-induced hearing loss.

The Noise at Work Regulations 1989 have set down more specific requirements on controlling noise and employers must carry out Risk Assessments, eliminate noise at source or reduce it to tolerable and 'safe' levels.

Failure to protect employees' hearing is an offence and carries fines in both the magistrates' or Crown Courts.

What are the hazards from noise?

Hearing damage

Exposure to high noise levels can cause incurable hearing damage. Usually, the important factors are:

- the noise level, given in decibel units as dB(A)
- the length of time over which people are exposed to the noise: daily and over a number of years.

Sometimes the peak pressure of the sound wave may be so great that there is a risk of instantaneous damage. This is most likely when explosive sources are involved, such as in cartridge-operated tools or guns.

The damage involves loss of hearing ability, possibly made worse by permanent tinnitus and other effects. Sufferers find it hard to distinguish words clearly, e.g. they tend to confuse words such as 'bit' and 'tip'.

Other effects of noise at work

Noise at work can cause other problems, such as disturbance, interference with communications and stress. Although the Noise at Work Regulations 1989 do not deal with these specifically, you should bear in mind that they might also need to be tackled.

What do employers have to do about noise?

Noise must be assessed to establish whether employees are subjected to unacceptable levels which could cause permanent hearing damage.

The persons carrying out noise assessments must be competent to do so. Environmental assessments may be appropriate but so might personal dose meters to establish exactly what each individual is exposed to.

Noise of all descriptions must be assessed, as must the cumulative effects of noise from numerous sources.

Is there anything I must know before undertaking a noise assessment?

Yes. An employer must be fully conversant with the Noise at Work Regulations and must understand the different noise levels at which action must be taken. These are called *action levels*.

Employers have a duty to reduce noise levels to the lowest practicable level.

Noise levels are calculated on the sound emitted from machines, processes, etc. and the length of time employees are exposed. Individuals can be exposed to high sound levels for short periods of time and not suffer hearing damage. The longer the exposure time to noise, the greater the risk of hearing damage.

Noise levels are measured in decibels (dB(A)). If the noise level in the workplace is below 85 dB(A) then the employer does not have to do anything specific to control the noise, merely be sure that the noise is as low as it can get.

If noise levels are above 85 dB(A) but below 90 dB(A), the employer has to carry out a noise risk assessment. Assessments must be done by competent persons. At this level, employees may request hearing defenders and the employer must provide them.

If noise is above 90 dB(A), a hearing protection zone must be declared and hearing defenders provided to all employees and visitors. Suitable safety signs must be displayed in hearing protection zones.

More stringent requirements are imposed on higher noise levels.

Are these noise action levels the permanent levels or will they change?

A new EU Noise Directive was tabled at various European Council meetings during 2000/2001 and, in November 2001, a new directive on noise was adopted by member states. This directive will repeal the earlier directive on which the Noise at Work Regulations 1989 were based. The UK will need to introduce new Noise at Work Regulations and this will have to be achieved by 2006.

The proposed Regulations will introduce new 'action levels' as follows:

(1) provide information and training to workers at 80 dB(A) (currently 85 dB(A))
(2) workers will have the right to hearing checks/audiometric testing at 85 dB(A) (as now) and also at 80 dB(A) as the risk is indicated
(3) make hearing protection available at 80 dB(A) (currently 85 dB(A))
(4) hearing protection to be worn at 85 dB(A) (currently 90 dB(A))
(5) limit on exposure to noise to be 87 dB(A) (currently no limit)
(6) programme of control measures at 85 dB(A) (currently 90 dB(A))
(7) designate noise control areas, display notices, etc. at 85 dB(A) (currently 90 dB(A))
(8) noise exposure which varies daily can be averaged over a week (currently eight hours).

Action levels are therefore going to reduce. This may seem quite reasonable — only a 5 dB(A) reduction. But, in noise level terms, a 3 dB(A) rise in noise level is, in effect, a doubling of the sound pressure levels. In simple terms, a lowering of noise action levels to 80 and 85 dB(A) will be quite significant.

How should a noise assessment be carried out?

Decide whether you might have a problem

If people have to shout or have difficulty being understood by someone about two metres away, you might have a problem. To be sure about this you will need to get the noise assessed.

Get the noise assessed

Your assessment should find out whether noise exposure is likely to reach the action levels. Where the assessment shows you have a noise problem, you should use it to help you develop plans for controlling exposure.

The job must be done by a competent person, someone who understands the Health and Safety Executive's guidance on assessment and how to apply it in the workplace. The essential qualification for the person is the ability to do the job properly and to know his or her own limits; this is more important than formal qualifications. However, many technicians may need extra training and local technical colleges often provide short courses lasting a few days or can advise on where they are available. Alternatively you might call a consultant.

Tell the workers affected

Where your assessment shows exposure is at or above any of the action levels, you should let employees know there is a noise hazard and what you want them to do to keep risks to a minimum.

Reduce the noise as far as reasonable practicable

Where the exposure needs to be controlled, the most reliable way is to quieten the workplace if this can be done.

You can avoid problems if you can make sure that noise reduction is built into new machinery when you buy it. Ask about noise before deciding which machine to buy.

You should also consider whether it might be possible to reduce either the number of people working in noisy areas or the time they have to spend in the areas. Perhaps some jobs can be done in a quieter location.

Ear protection

If people have to work in noise-hazardous areas, they will need ear protectors (ear muffs or ear plugs). However, these should not be regarded as a substitute for noise reduction. As long as people work in noise at or above the second or the peak action level, the Regulations still require you to reduce the noise exposure by other means as far as this is reasonably practicable.

Between the first and second action levels, i.e. between 85 dB(A) and 90 dB(A), you should make sure that:

- protection is freely available
- the employees know that unless they wear it, there is some risk to their hearing.

The Regulations do not, however, make it a legal duty for employees to wear protection below the second action level.

Make sure that young people in particular get into the routine of wearing ear protectors before their hearing is damaged.

Where use of protection is compulsory, ear protection zones should be marked if this is reasonably practicable. Ensure that everyone who goes into a marked zone, even for a short time, uses ear protection.

Check to make sure your programme is working

Make sure the equipment you provide is kept in good condition.

If you rely on ear protectors, find out whether they are really being used. If anything is wrong do not neglect it, put it right!

What information should be given to employees?

Employees must, by law, be provided with information about any risk to their hearing, especially if they will be exposed to levels of 85 dB(A) or above.

Adequate information, instruction and training is required so that the employee understands:

- the risk of damage to their hearing
- the steps which they can take to minimise the risk
- the procedure they need to follow to obtain personal ear protection
- their own duties under the Regulations.

What types of ear protection are available?

There are generally two main styles or types of ear defender:

- the 'headphone' type which cover the ears completely
- ear plugs which fit into the ear canal.

Ear protectors will only be effective if they are in good condition and properly maintained.

They must suit the individual and be worn properly. It is vitally important therefore to consult employees on what types they prefer — it should be an individual choice.

Do employers have other responsibilities besides providing ear protection?

Yes. The main responsibility of employers is to reduce the noise *at source* to the lowest level possible.

Noise can be controlled by:

Engineering controls	purchasing equipment with low noise emissions changing the process, e.g. presses instead of hammers avoiding metal to metal impacts using flexible couplings and mountings introducing design dampers
Orientation and location	correct sizing of ductwork, fans, motors, etc. move the noise source away from employees, turning machines around so that noise or sound waves can travel out of the building not putting machines, etc. into hard surface areas as noise 'bounces' off surfaces
Enclosure	surround the machine or noise source in sound-absorbing material (total enclosure is most effective) soundproof the room/work area introduce sound-absorbent materials to surfaces
Use of silencers	use on ductwork, for motors use on pipes which carry gas, air or steam use on exhaust ventilation systems

Lagging	lag pipes as an alternative to enclosure
Damping	dampers can be fitted to ductwork use double skin design, preferably with noise-absorbent material in between
Absorption	acoustic ceiling and wall panels help to absorb the sound waves
Screens	temporary acoustic screens can help to reduce levels of noise and these can be moved to where needed
Isolate workers	remove workers from the noise source by construction acoustic booths for them to work in

Usually it will need specialist noise or acoustic consultants to work out exactly what needs to be done to reduce noise to safe levels. Remember, noise must be a combination of different machines.

If you feel that you have 'a din' in the workplace, then you will need expert advice to reduce noise levels to tolerable levels.

9

Display screen equipment

What is the main legislation which relates to health and safety issues in respect of display screen equipment?

The Health and Safety (Display Screen Equipment) Regulations 1992 came into force in 1993 along with the other 'six pack' Health and Safety Regulations. They were slightly amended in 2002.

The main points of the Display Screen Equipment Regulations are:

- employers must assess all relevant work stations, i.e. carry out Risk Assessments
- employers must reduce all risks to the lowest extent reasonably practicable
- workstations must meet minimum requirements as contained in the Schedule to the Regulations covering:
  - display screen
  - keyboard
  - work desk or surface
  - work chair
  - space requirements
  - lighting
  - reflection and glare

- ○ noise
- ○ heat
- ○ radiation
- ○ humidity
- ○ computer/user operator interface.

Why do we need to worry about the health and safety aspects of using display screens, keyboards, etc.?

There are health problems associated with the frequent use of display screen equipment or visual display units.

The health effects may not be immediately obvious but severe disability can ensue with repeated use of the equipment and the exacerbation of the symptoms.

Health problems associated with display screens or VDUs are:

- upper limb disorders, including pains in the neck, arms, elbows, wrists, hands and fingers — these injuries are often collectively refereed to as 'repetitive strain injuries'
- backache
- fatigue and stress
- temporary eye strain but not eye damage
- headaches.

The health symptoms associated with the use of display screens are not always the direct result of one thing — it is usual for a combination of factors to occur to create the symptoms. A failure to eliminate the hazards means that people will repeatedly suffer and the effects of the hazard will accumulate to cause chronic symptoms.

Do the Regulations affect everybody who uses display screen equipment?

No. The Regulations in the main apply to the *users* of display screen equipment. The Regulations also only apply to employees and the self-employed. Therefore a 'user' can only be an employee or a self-employed person.

Regulation 1(2d) defines a 'user' as an employee who habitually uses display screen equipment as a significant part of his/her normal work.

In the same Regulation, a self-employed person who habitually uses display screen equipment is defined as an 'operator'.

A display screen is not only a computer screen but also television screens, video screens, plasma screens, microfiche screens, etc. Emergency technologies are creating new types of screens and it is anticipated that the Regulations will cover all of these.

Employers must decide who is a user or an operator under the Regulations and apply the requirements of the relevant regulations.

Workers who do not input or extract information from a display screen are generally not users.

Employers need to ask themselves a few searching questions in order to ascertain whether they have users or operators on their staff.

First question: Do any of my employees or the people whom I engage as 'contractors' normally use display screen equipment (DSE) for continuous or near-continuous spells of an hour or more at a time?

Second question: Do any of them use DSE for an hour or more, more or less daily?

Third question: Do they have to transfer information quickly to or from the DSE?

Fourth question: Do they need to apply high levels of attention and concentration to the work that they do?

Fifth question:	Are they highly dependent on their DSE to do their job or have they little choice in using it?
Sixth question:	Do they need special training or skills to use the DSE?

Part-time or flexible workers must be assessed on the same criteria because it is not the length of time they spend 'at work' which counts but the length of time they spend using the VDU/DSE.

Sometimes, employers may wish to simplify things and class *all* users of display screen equipment as 'users' or 'operators' under the terms of the Regulations. This means that the good practice requirements of the Regulation will be applied throughout the organisation.

Examples of display screen users

- Typist, secretary, administration assistant who uses a PC or word processor for typing documents, etc.
- Word processing worker
- Data entry clerk/operator
- Database operator and/or creator
- Telesales personnel
- Customer service personnel if computer entry of information is a common part of the job
- Journalists, editorial writers
- TV/video editing technicians
- Micro-electronics testing operators who use DSE to view test results, etc.
- CAD technicians
- Air traffic controllers
- Graphic artists
- Financial dealers

Is it easy to define who is not a user of display screen equipment?

Yes, relatively so.

The answers to the six questions listed previously should enable the employer to easily differentiate who is who.

Anyone who uses display screen equipment occasionally will not be a user under the Regulations. Nor will anyone who can choose when or for how long they use DSE.

Laptop users will probably *not* be users as they can (usually) choose when, where and for how long they use their screens and computer.

Receptionists will often not be classed as users as they are not continuously using their screens (unless they predominately operate a switchboard which relies on a screen for extension transfers, etc.).

Are employees who work at home covered by the Regulations?

If they are an employee and they use their display screen equipment continuously as part of their job, they will be defined as a user irrespective of where they use the equipment.

The display screen and workstation does not have to be supplied by the employer — an employee can provide their own equipment but the employer would still have to comply with their duties in respect of users and operators and assess the hazards and risks to health.

In order to determine whether homeworkers are users or operators of display screen equipment the six questions posed earlier will need to be asked of each individual worker.

What is a 'workstation' and how do the Regulations apply to these?

A workstation is defined in Regulation 1 of the Health and Safety (Display Screen Equipment) Regulations 1992 as:

> an assembly comprising:
> (i) display screen equipment
> (ii) keyboard or other input device
> (iii) optional software
> (iv) optional accessories to the display screen equipment
> (v) any disk drive, telephone, modem, printer, document holder, work chair, work desk, work surface or other item peripheral to the display screen equipment
> (vi) the immediate work environment around the display screen equipment.

Regulation 2 requires employers to perform a suitable and sufficient analysis of workstations which:

- are used for the purposes of this, his undertaking (regardless of who provided them), by users
- have been provided by him and are used by operators

in order to assess the health and safety risks to which those people are exposed as a result of that use.

The analysis is to assess and reduce risk — it is a Risk Assessment.

Is it necessary to complete a Risk Assessment for each workstation and user or operator?

Yes, because each person is different and the effect that using the display screen *may* have on them will be different for individuals. Of course, some individuals may have no ill-effects from using a display screen and there will be little you will need to do.

Individual workstations may vary in design, people's tasks will be different, the amount of control they have over their jobs may be different.

The most effective way to conduct Risk Assessments for display screen equipment users is to create a questionnaire which includes all the relevant sections on:

- display screens
- keyboards
- mouse or trackball
- software
- furniture
- environment.

The workstation analysis or Risk Assessment is best done by the individual concerned, once they have had proper training in what they are to look for and how to record the information.

Do the Regulations only require an employer to carry out these Risk Assessments?

No, they are one part of the employer's responsibilities under the Regulations.

The Regulations themselves require employers to:

- analyse workstations to assess and reduce risks
- ensure that workstations meet minimum specified requirements
- plan work activities so that they include short breaks or changes of activity
- provide eye and eyesight tests on request and special spectacles if needed
- provide information and training.

What are the 'minimum specified requirements' for workstations?

The Regulations are quite specific about requirements for workstations and the appropriate Regulation, Regulation 3, was amended in 2002 to address an European Ruling on the interpretation of the Regulation applying to workstations.

The 'minimum specified requirements' apply to all workstations provided by an employer and not just to those used by 'users or operators'.

The European Court, in effect, stated that all workers, employees or others who used a workstation while at work were entitled to have a workstation which met the 'minimum specified requirement'.

Do all workstations have to be modified to meet these requirements?

If workstations do not already comply they will need to be modified to meet the conditions laid out in paragraph 1 of Schedule 1 to the DSE Regulations.

Paragraph 1 of the Schedule lists the following requirements:

- the components required, e.g. document holder, chair, desk, etc. are present at the workstation
- they relate to health, safety and welfare
- the inherent requirements or characteristic of the task make compliance appropriate.

In effect, employers have to ensure that all workers using DSE have a suitable environment in which to work, have the necessary equipment to work safely and that the tasks they do are managed effectively so as not to create health and safety issues.

What are the main areas to pay attention to when carrying out a workstation assessment or Risk Assessment?

Each workstation should be assessed with the following in mind:

- adequate lighting
- adequate contrast — no glare or distracting reflections
- distracting noise minimised
- leg room and clearances to allow postural changes
- window covering if needed to minimise glare
- software — appropriate to the task, adapted to the user, no undisclosed monitoring of the user
- screen — stable image, adjustable, readable, glare- and reflection-free
- keyboard — usable, adjustable, detachable, legible
- work surface with space for flexible arrangement of equipment and documents, glare-free
- chair — stable and adjustable
- footrest, arm/wrist rest if users need one.

Are all display screen equipment users entitled to an eyesight test and a free pair of glasses?

No. Only those employees who are classed as 'users' under the Regulations are covered by the Regulation applying to eyesight tests.

An employee who is a user of DSE can request an eyesight test, as can anyone who is to *become* a user, and the employer has to arrange for one to be carried out.

If an existing user requests a test, an employer must arrange for it to be carried out as soon as practicable after the request and for a potential user, before they become a user.

The continual use of DSE or VDU screens may cause visual fatigue and headaches and corrective glasses may reduce the eye strain often experienced. There is no evidence yet available, however, that frequent use of display screen equipment causes permanent eye damage or creates poor eyesight. Users with pre-existing sight conditions may just become a little more aware of them.

Eyesight tests should be carried out by competent professionals and must consider the effects of working at DSE so the optician (or medical equivalent) will need to know that the eyesight test is for working at display screens, etc.

Once an existing user has had an eyesight test, they can request one at regular intervals. The employer should determine what this interval should be with the user of the equipment and should take advice from the optician or other expert.

Eyesight tests which detect short or long sight, eye defects, etc. are *not* the responsibility of the employer — they need only concern themselves with an eye test which addresses any safety or health issues with regaed to using display screens.

The employer must arrange for an eyesight test when requested to do so. This could be by having arrangements with local opticians or by having the eye testing carried out on the premises by mobile health surveillance units, etc.

The employer can make arrangements with only one local optician and employees will have no choice who they visit. Alternatively, employers can have employees use their own optician if they prefer. The important thing for the employer is that they must facilitate such eye sight tests if requested to do so.

Employers are not responsible for the costs of 'normal' corrective spectacles — these are at the employees'/users' own expense. But an employer is responsible for the cost of any 'special' corrective appliances which the optician has determined need to be worn by the user to prevent them suffering unnecessary eye strain while using a display screen. The user is only entitled to a basic pair of corrective spectacles necessary for them to continue to use the display screen safely. 'Designer' frames, special lenses, etc. are not the responsibility of the employer.

Employers may make a contribution towards the cost of other types of corrective spectacles if those spectacles include the 'special corrective' features needed for the DSE work.

What does the employer need to do in respect of the provision of training regarding the use of display screen equipment?

The Regulations are quite specific about the duties of employers to provide training and information to display screen users.

The employer has to ensure that 'users' and those about to become 'users' of display screen equipment receive adequate health and safety training in the use of any workstation on which they may be required to work.

Training should be provided before a new employee becomes a user of the equipment. The purpose of the training is to ensure that those who are (or will be) users know and understand the hazards and risk associated with using display screen equipment.

Training on DSE can be incorporated into general health and safety training or induction programmes as it is good practice for everyone to be aware of hazards and risks of all work activities.

It is important that any training programme addresses the steps needed to reduce or minimise the following risks:

- musculo-skeletal problems
- visual fatigue
- mental stress.

Managers of those using DSE also need to be trained in health and safety issues relating to DSE as they can have important influence over a key health hazard — mental stress. Managers must be aware of the legal need for users to take regular breaks away from the screen, for workstations to be ergonomically friendly, etc.

Users must be trained on how to use their display screen equipment effectively. They must know how to make their own personal adjustments to height, tilt of screen, contrast, etc. They must know when they can take breaks and what other tasks are expected.

As with all health and safety training, it is important for employers to have a record-keeping system so that they will be able to demonstrate, if called upon to do so, that their employees, users or operators received suitable and sufficient training.

Occupational health issues can take several years to manifest themselves and elements of musculo-skeletal injury may occur after an employee, user or operator has left the company.

Evidence of training can be useful to show that, as an employer, you fulfilled your statutory duties and that the employee or user was aware of the hazards and risks and knew what to do to control them.

Keep training records for at least six years — longer is preferable. Computerised records must comply with the Data Protection Act 1998.

What information do users have to be provided with?

Users of display screen equipment must be provided with adequate information about:

- all aspects of health and safety relating to their workstations
- the steps taken by their employer to ensure compliance with the Regulations.

Users and operators of DSE need to know about the Risk Assessments which have been undertaken, the hazards and risks identified and the control measures that the employer has put in place to reduce the hazards and risks.

In addition, users and operators need to know what procedures are in place for them to have eye-sight tests, the frequency of tests, the provisions for the purchase of 'special needs' spectacles, etc.

Information should be given on when breaks can be taken, what other tasks need to be completed during these times, when they should have training, etc.

Employers should not forget their general duties to *all* employees and others, in respect of information, instruction and training on all work activities as required under the Management of Health and Safety at Work Regulations 1999.

Top tips

- Assess all workstations in the organisation.
- Review all equipment for comfort, ease of use etc.:
  - chairs
  - screens
  - keyboard
  - mouse/trackball
  - lighting/glare
  - environment
  - work/schedule demands
  - software.
- Record information in Risk Assessments.
- Decide on control measures to reduce hazards and risks.
- Offer eyesight tests to users — do not wait for them to ask.
- Determine what level of corrective spectacle you will pay for or towards.
- Introduce a comprehensive training programme and give out good levels of information.
- Keep training records for as long as possible.

10

Workplace facilities

The Workplace (Health, Safety and Welfare) Regulations 1992 set out the duties of employers in respect of providing their employees with suitable and safe working conditions and facilities.

The Regulations aim to ensure that workplaces meet the health, safety and welfare needs of all members of the workforce, including people with disabilities. Several of the Regulations require things to be 'suitable'. Regulation 2(3) makes it clear that things should be suitable for anyone. This includes people with disabilities. When the workforce includes people with disabilities, it is important to ensure that the workplace is suitable for them, particularly traffic routes, toilets and workstations.

Definitions

'Workplace' — these Regulations apply to a very wide range of workplaces, including places of entertainment. The term also includes the common parts of shared buildings, private roads and paths on industrial estates and business parks, and temporary worksites (but not construction sites).

'Work' means work as an employee or self-employed person.

'Premises' means any place including an outdoor place.

'Domestic premises' means a private dwelling. These Regulations do not apply to domestic premises, and exclude homeworkers. However, they do apply to hotels, nursing homes and to parts of workplaces where 'domestic' staff are employed, such as the kitchens of hostels.

What are some of the specific requirements of the Regulations?

Health issues

Ventilation

Workplaces need to be adequately ventilated. Fresh, clean air should be drawn from a source outside the workplace, uncontaminated by discharges from flues, chimneys or other process outlets, and be circulated through the workrooms.

Ventilation should also remove and dilute warm, humid air and provide air movement which gives a sense of freshness without causing a draught. If the workplace contains process or heating equipment or other sources of dust, fumes or vapours, more fresh air will be necessary to provide adequate ventilation.

Windows or other openings may provide sufficient ventilation but, where necessary, mechanical ventilation systems should be provided and regularly maintained.

These Regulations do not prevent the use of unflued heating systems designed and installed to be used without a conventional flue.

Any mechanical ventilation, including air conditioning systems, should be regularly and properly cleaned, tested and maintained to ensure that they are kept clean and free from anything which may contaminate the air.

If ventilation systems are required to provide dilution for obnoxious fumes, etc. they may need a breakdown warning device fitted.

If ventilation has to be kept to a minimum for the process purpose, then employees must be given regular breaks from the environment, e.g. mushroom growing greenhouses.

Temperatures in indoor workplaces

Comfort depends on air temperature, radiant heat, air movement and humidity. Individual personal preference makes it difficult to specify a thermal environment which satisfies everyone.

For workplaces where the activity is mainly sedentary, for example offices, the temperature should normally be at least 16°C. If work involves physical effort, it should be at least 13°C (unless other laws require lower temperatures, e.g. food hygiene legislation).

Work in hot or cold environments

The risk to the health of workers increases as conditions move further away from those generally accepted as comfortable. Risk of heat stress arises, for example, from working in high air temperatures, exposure to thermal radiation or high levels of humidity, such as those found in foundries, glassworks and laundries. Cold stress may arise, for example, from working in cold stores, food preparation areas and in the open air during winter.

Assessment of risk to workers' health from working in either a hot or cold environment should consider two sets of factors — personal and environmental. Personal factors include body activity, the amount and type of clothing and duration of exposure. Environmental factors include ambient temperature and radiant heat and, if the work is outside, sunlight, wind velocity and presence of rain or snow.

Any assessment should consider the following.

- Measures to control the workplace environment, in particular heat from any source. Minimising the risk of heat stress may mean insulating plant which acts as a source of radiant heat, using local cooling by increasing ventilation rates and maintaining the appropriate level of humidity. If it is not reasonably practicable to avoid workers being exposed to cold environments you should consider using local environmental controls, e.g. cab heaters in fork-lift trucks used in cold stores.

- Restriction of exposure by, for example, reorganising tasks to build in rest periods or other breaks from work. This will allow workers to rest in an area where the environment is comfortable and, if necessary, to replace bodily fluids to combat dehydration or cold. If work rates cause sweating, workers may need frequent rest pauses for changing into dry clothing.
- Medical preselection of employees to ensure they are fit to work in these environments.
- Use of suitable clothing (which may need to be heat-resistant or insulating, depending on whether the risk is from heat or cold).
- Acclimatisation of employees to the environment in which they work.
- Training in the precautions to be taken.
- Supervision to ensure that the precautions identified by the assessment are taken.

Lighting

Lighting should be sufficient to enable employees to work and move about safely. If necessary, local lighting should be provided at individual workstations, and at places of particular risk, such as crossing points on traffic routes. Lighting and light fittings should not create any hazard.

Automatic emergency lighting, powered by an independent source, should be provided where sudden loss of light would create a risk.

Dazzling light and annoying glare should be avoided, as should coloured lights which cast shadows.

Staircases and emergency exit routes must always be adequatly lit.

Lights should be replaced, repaired and cleared as necessary.

Natural lighting should be encouraged whenever possible as it has a beneficial psychological effect on employees.

Cleanliness and waste materials

Every workplace and the furniture, furnishings and fittings should be kept clean and it should be possible to keep the surfaces of floors, walls and ceilings clean. Cleaning and the removal of waste should be carried out as necessary by an effective method. Waste should be stored in suitable receptacles.

Room dimensions and space

Workrooms should have enough free space to allow people to move about with ease. The volume of the room, when empty, divided by the number of people normally working in it should be at least 11 metres per person. This is a minimum and may be insufficient depending on the layout, contents and the nature of the work.

Workstations and seating

Workstations should be suitable for the people using them and for the work. Ergonomic assessments should be undertaken to ensure suitability and should consider the size of the individual and the work to be undertaken. People should be able to leave workstations swiftly in an emergency. If work can or must be done sitting, seats which are suitable for the people using them and for the work they do they should be provided. Seating should give adequate support for the lower back, should be adjustable, stable and comfortable, and footrests should be provided for workers who cannot place their feet flat on the floor.

Safety issues

Maintenance

The workplace and certain equipment, devices and systems should be maintained in efficient working order (efficient for health, safety and welfare). Such maintenance is required for mechanical ventilation systems, equipment and devices which would cause a risk to health and safety or welfare should a fault occur.

Floors and traffic routes

'Traffic route' means a route for pedestrian traffic, vehicles, or both, and includes any stairs, fixed ladders, doorways, gateways, loading bays or ramps. There should be sufficient traffic routes, of sufficient width and headroom, to allow people and vehicles to circulate safely with ease.

Floors and traffic routes should be sound and strong enough for the loads placed on them and the traffic expected to use them. The surfaces should not have holes, be uneven or slippery and should be kept free of obstructions.

Restrictions should be clearly indicated. Where sharp or blind bends are unavoidable or vehicles need to reverse, measures such as one-way systems and visibility mirrors should be considered. Speed limits should be set. Screens should be provided to protect people who work where they would be at risk from exhaust fumes, or to protect people from materials likely to fall from vehicles.

Additional measures need to be taken where pedestrians have to cross or share vehicle routes. These may include marking of routes, provision of crossing points, bridges, subways and barriers.

Open sides of staircases should be fenced with an upper rail at 900 mm or higher and a lower rail. A handrail should be provided on at least one side of every staircase and on both sides if there is a particular risk. Additional handrails may be required down the centre of wide staircases. Access between floors should not be by ladders or steep stairs.

Falls and falling objects

The consequences of falling from heights or into dangerous substances are so serious that a high standard of protection is required. Secure fencing should be provided to prevent people falling from edges and objects falling onto people. Where fencing cannot be provided, other measures should be taken to prevent falls.

If a person might fall two metres or more, or might fall less than two metres and risk serous injury, fencing should be at least 1100 mm high and have two guard-rails. Tanks, pits and structures should be securely covered or fenced to a height of a least 1100 mm.

Fixed ladders should be of sound construction, properly maintained and securely fixed. Rungs should be horizontal and give adequate foothold and the stiles should extend at least 1100 mm above the landing. Fixed ladders over 2.5 metres in length at a pitch of more than 75° should be fitted with safety hoops or permanently fixed fall-arrest systems. Fixed ladders should only be used if it is not practical to install a staircase.

Slips and trips which may be trivial at ground level may result in fatal accidents when on roofs. Precautions should be taken when there is a risk of falling off or through a roof. These may include fall-arrest devices and crawling boards. Fragile roofs or surfaces should be clearly identified.

If fencing or covers cannot be provided or have to be removed, effective measures should be taken to prevent falls. Access should be limited to specified people and, in high risk situations, suitable formal written Permit to Work systems should be adopted.

A safe system of work should be operated which may include the use of a fall-arrest system or safety lines and harnesses and secure anchorage points. Systems which do not require disconnection and reconnection of safety harness should be used. If there is no need to approach edges, the length of the line and anchorage position should prevent the edge being approached.

Materials and objects must be stacked and stored in such a way that they are not likely to fall and cause injury. Storage racking and shelving must be of adequate strength and stability for the loads to be placed on it. In general, racking and shelving is made from lightweight materials and is limited to the amount of wear and tear it can withstand. The skill of workplace transport operators has a great bearing on the amount of damage likely to be caused. The greater the damage to racking and shelving, the weaker it will be until it may eventually collapse, even when supporting less than its normal load.

To ensure that racking or shelving installations continue to be serviceable:

- check regularly to identify damage and necessary actions
- encourage employees to report any damage, however minor, so that its effect on safety may be assessed

- fix maximum load notices and adhere to them strictly.

Appropriate precautions in stacking and storing include:

- safe stacking in sound pallets
- banding or wrapping to prevent individual articles falling
- setting limits for the height of stacks to maintain stability
- regular inspection of stacks to detect and remedy any unsafe stacks
- instruction and training of employees in stacking
- special arrangements for objects which may be difficult to store.

Transparent and translucent doors, gates or walls and windows

Windows, transparent or translucent surfaces in walls, partitions, door and gates should, where necessary for reasons of health and safety, be made of safety material or be protected against breakage. If there is a danger of people coming into contact with these surfaces, they should be marked or incorporate features to make this apparent.

Employers will need to consider whether there is a foreseeable risk of people coming into contact with glazing and being hurt. If this is the case, the glazing will need to meet the requirements of the Regulations.

Openable windows and the ability to clean them safely

Openable windows, skylights and ventilators should be capable of being opened, closed or adjusted safely and, when open, should not be dangerous.

Windows and skylights should be designed so that they may be cleaned safely. When considering if they can be cleaned safely, account may be taken of equipment used in conjunction with the window or skylight or of devices fitted to the building.

Doors and gates

Doors and gates should be suitably constructed and fitted with safety devices (e.g. panic bolts — a door bolt that can be operated from the inside in an emergency) if necessary.

Doors and gates which swing both ways and conventionally hinged doors on main traffic routes should have a transparent viewing panel.

Power-operated doors and gates should have safety features to prevent people being struck or trapped and, where necessary, should have a readily identifiable and accessible control switch or device so that they can be stopped quickly in an emergency.

Upward-opening doors or gates need to be fitted with an effective device to prevent them falling back. Provided that they are properly maintained, counterbalance springs and similar counterbalance or ratchet devices to hold them in the open position are acceptable.

Escalators and moving walkways

Escalators and moving walkways should function safely, be equipped with any necessary safety devices, and be fitted with one or more emergency stop controls which are easily identifiable and readily accessible.

Welfare issues

Sanitary conveniences and washing facilities

Suitable and sufficient sanitary conveniences and washing facilities should be provided at readily accessible places. They and the rooms containing them should be kept clean and be adequately ventilated and lit. Washing facilities should have running hot and cold or warm water, soap and clean towels or other means of cleaning and drying. If required by the type of work, showers should also be available. Men and women should have separate facilities unless each facility is in a separate room with a lockable door and is for use by one person at a time.

Drinking water

An adequate supply of wholesome drinking water, with an upward jet or suitable cups should be provided. Water should only be provided in refillable enclosed containers where it cannot be obtained directly from a mains supply. The containers should be refilled at least daily (unless the are chilled water dispensers where the containers are returned to the supplier for refilling). Bottled water/water dispensing systems may still be provided as a secondary source of drinking water.

Accommodation for clothing and facilities for changing

Adequate, suitable and secure space should be provided to store workers' own clothing and special clothing. As far as is reasonable practicable, the facilities should allow for drying clothes. Changing facilities should also be provided for workers who change into special work clothing. The facilities should be readily accessible from workrooms and washing and eating facilities and should ensure the privacy of the user.

Facilities for rest and to eat meals

Suitable, sufficient and readily accessible rest facilities should be provided. Rest areas or rooms should be large enough and have sufficient seats with backrests and tables for the number of workers likely to use them at any time. They should include suitable facilities to eat meals where meals are regularly eaten in the workplace and the food would otherwise be likely to be contaminated.

Seats should be provided for employees to use during their breaks. These should be in a place where personal protective equipment need not be worn. Work areas can be counted as rest areas and as eating facilities providing they are adequately clean and there is a suitable surface on which to place food. Where provided, eating facilities should include a facility for preparing or obtaining a hot drink. Where hot food cannot be obtained in or reasonable near to the workplace, employees may need to be provided with a means for heating their own food.

Canteens or restaurants may be used as rest facilities, provided there is no obligation to purchase food.

Suitable rest facilities should be provided for pregnant women and nursing mothers. They should be near to sanitary facilities and, where necessary, include the facility to lie down.

Rest areas should be away from the workstation and should include suitable arrangements to protect non-smokers from discomfort caused by tobacco smoke.

Does an employer have to legally provide employees with showers?

No, unless the nature of the work requires them, or they are needed for health reasons.

Employees who are exposed to dusty atmospheres may require showers to ensure that they decontaminate themselves before leaving the workplace.

Asbestos workers are required by law to be given showering facilities (under the Control of Asbestos at Work Regulations 2002).

Catering staff, or kitchen workers may need to have showers provided because of the hot, humid atmosphere of the kitchen or because showering helps to reduce surface body bacteria which in turn will reduce the risk of food contamination.

Remember, however, that the law is a minimum set of standards, and additional welfare and washing facilities may prove a good investment in respect of employee morale, etc.

What are the recommended numbers of water closets, urinals and wash hand-basins which need to be provided for employees?

The 'Workplace' Regulations only stipulate that 'suitable and sufficient' facilities should be provided.

The guidance in the Approved Code of Practice gives an indication of suitable and sufficient as follows.

No. of people at work (male and female)	No. of water closets	No. of wash stations
1–5	1	1
6–25	2	2
26–50	3	3
51–75	4	4
76–100	5	5

If accommodation is to be provided for men only, then water closets and urinals can be provided as follows.

No. of men at work	No. of water closets	No. of urinals
1–15	1	1
16–30	2	1
31–45	2	2
46–60	3	2
61–75	3	3
76–90	4	3
91–100	4	4

Wash hand-basins should be provided for every water closet plus reasonable additions for the urinals.

Does an employer legally have to provide sanitary accommodation for members of the public?

No, not under Health and Safety Law but employers may need to under other legislation, particularly if the business is one which sells food and drink to members of the public.

If employers do provide publicly accessible sanitary accommodation, they must make sure that their employees are not prejudiced because of lack of facilities. Employees should be able to use the facilities 'without undue delay'.

Checklist

The Workplace (Health, Safety and Welfare) Regulations 1992

All places of work where employees work must comply with applicable requirements, e.g. those detailed below.

- Workplaces to be kept clean, well maintained, in good order and repair. Equipment or premises which could fail must be subject to adequate maintenance procedures.
- Ventilation must be suitable and sufficient, incorporating fresh or purified air. Any mechanical ventilation systems must be adequately maintained.
- Reasonable temperatures must be maintained during working hours inside all buildings. No injurious or offensive fumes may enter the workplace. Suitable thermometers to be provided. Recommended temperature for sedentary work is 16°C *but* this is only advisory.
- Suitable and sufficient lighting is required and, where artificial light is provided, if it should fail, emergency lighting is required.
- Walls, floors and ceiling surfaces must be capable of being kept clean. Furniture and fixtures, etc. must be kept clean.
- Persons at work to have adequate working space and must not be overcrowded.
- Workstations must be suitably positioned and, if outside, suitably protected from the weather.
- Arrangements must be in place for leaving the workstation in an emergency.

- Floors and traffic routes must be constructed in a suitable manner for the purpose, i.e. having no holes or slopes.
- Suitable precautions must be taken to protect people falling from heights.
- Windows, doors and any panels, etc. glazed wholly or partially with translucent glass must be of safe material and adequately marked. Risk Assessment on glazing is required.
- Windows and skylights should be designed so that, when opened, closed or adjusted, they cause no danger to anyone.
- Safe circulation of pedestrians and vehicles must be organised in the workplace. Traffic routes are required.
- Doors and gates must be fitted with any necessary safety devices, e.g. to sliding doors.
- Escalators must function safely.
- Sanitary accommodation must be suitable and sufficient and readily accessible to all persons at work.
- Washing facilities must be provided together with adequate soap, towels, hot water, etc.
- Wholesome drinking water must be provided.
- Suitable accommodation for outdoor clothing is required, and suitable changing facilities must be provided.
- Rest facilities are to be provided, including facilities to eat meals and take breaks.
- Rest rooms must protect non-smokers from tobacco smoke.
- Suitable facilities must be provided for pregnant women or nursing mothers.

11

Smoking at work

Does the law prohibit smoking at work?

No, the Health and Safety at Work Etc. Act 1974 does not ban smoking at work, neither do Regulations made under the Act.

The law requires employers to safeguard the health, safety and welfare of their employees while they are at work.

Section 2(1) of the Health and Safety at Work Etc. Act 1974 could be used to require employers to take steps to protect their employees from the effects of smoking while they are at work. However, employers have to do what is 'reasonably practicable' in respect of managing workplace health and safety.

The Management of Health and Safety at Work Regulations 1999 require employers to carry out Risk Assessments of all potential health and safety hazards to which employees may be exposed. This will include the effect of smoking on those employees who do not smoke.

What does the law require in respect of smoking at work?

The Workplace (Health, Safety and Welfare) Regulations 1992 contain references to smoking at work in respect of the following:

- Regulation 25(3) requires employers to protect non-smokers from the discomfort of tobacco smoke in rest rooms and rest areas.

This means that employers cannot allow smokers and non-smokers to use the same rest room at the same time unless the room is effectively separated or ventilated.

Regulation 6 of the same Regulations requires employers to provide effective ventilation to workplaces and, if ventilation is insufficient, this could create a health hazard to employees.

The Approved Code of Practice which accompanies the Workplace (Health, Safety and Welfare) Regulations 1992 states that employers should either ban smoking in rest rooms, or provide separate rest rooms.

Another practical solution would be to allow non-smokers to take their breaks together and then for smoking employees to use the rest room/area.

What is the risk with regard to smoking at work and passive smoking?

People who smoke know the risks which they are exposing themselves to and so there is no additional risks to their health if they smoke at work, unless the work environment or use of chemicals, etc. increases or exacerbates the health risks.

There is a risk to people who do not smoke if they are exposed to 'environmental tobacco smoke' and persistent exposure can lead to lung cancer and other respiratory diseases.

Research in America indicates that passive smoking contributes to:

- lung cancer — long-term exposure of non-smokers to environmental tobacco smoke caused an increased risk of lung cancer, in the region of 20–30%

- heart disease — non-smokers in regular contact with smokers have an increased risk of heart disease of 30%
- stroke — passive smoking as well as active smoking increases the risk of stroke — sometimes by as much as 82%
- respiratory diseases — exposure to environmental tobacco smoke contributes to coughs, phlegm production, chest discomfort and reduced lung function. Those with asthma are especially affected by passive smoking
- nasal sinus cancer, childhood illnesses, e.g. leukaemia, meningitis, are also being investigated to establish whether such conditions are being exacerbated by exposure to passive smoking.

What are the steps to setting up a smoking policy for the work environment?

Generally, a five-step approach can be taken to establishing a smoking policy for the workplace.

- Step 1: set up a working party
    - include representatives from all work departments, sectors, etc. and include smokers, non-smokers and ex-smokers
    - review current practice — what happens about smoking in the workplace?
    - set out some basic objectives — what do you want to achieve?
    - what sort of complaints do employees raise? Who deals with them, how, etc.?
- Step 2: Raise the issue in the workplace
    - circulate information about health hazards from passive smoking and the options available to reduce item
    - use notice-boards, in-house journals, the intranet, emails, staff meeting, etc.
    - encourage feedback.

- Step 3: Consult the workforce
    - consultation is essential under the Consultation with Employees Regulations and any health and safety issues must be discussed with employees themselves or their representatives
    - consult any trade union representatives
    - explain that decisions have not yet been taken but this is a process of evaluating risks, seeking solutions, communicating, etc.
    - carry out a survey of employee concerns, those who smoke, etc.
    - what sort of solutions do employees want to see?
- Step 4: Formulate the policy
    - set out the principles of what needs to be achieved to safeguard the health of non-smoking employees
    - set out any legal requirements
    - set out what the hazards are
    - set out the options for change of policy, practice, etc.
    - set out what will be acceptable, e.g. smoking in nominated staff room only, whether 'smoking breaks' will be permissible, etc.
    - circulate the proposed policy for comment. Listen to the feedback and be prepared to make amendments to address any concerns.
- Step 5: Implement the policy
    - announce the final policy to the workforce
    - set a date for implementation — allow time for adjustment. Be flexible. People will have to change their habits — remember that smoking is addictive
    - twelve weeks is often recommended as the implementation period
    - check employment law for final details if necessary. Changing the smoking policy at work may be changing employees' contract terms and conditions
    - allocate resources to alter signage, etc., create new staff rooms, erect 'butt bins'

- inform everyone of final 'D-day' date and the consequences for non-compliance — with, perhaps, a lenient period for advice before any disciplinary process may be instigated
- supply help and information. Encourage smokers to give up — provide helpline and advice. Run support groups at lunch time, etc.
- ensure that all new employees know that the company has a no-smoking policy or restricted smoking policy, etc.

How would the law on passive smoking be enforced?

Generally, Section 2 of the Health and Safety at Work Etc. Act 1974 could be used when an enforcing officer believes that there is a health risk from passive smoking to employees.

Employers are required to take reasonably practicable steps to ensure the health, safety and welfare of their employees, and others, while they are at work.

However, the term 'reasonably practicable' would be quite important as the benefit of the improvement against the cost of implementing it would need to be considered.

Employers must complete a Risk Assessment under the Management of Health and Safety at Work Regulations 1999 on the effect of smoking at work on all employees. Obviously, the risk to those who already smoke will be minimal but the employer has to consider non-smokers.

Enforcement officers are encouraged to address the health hazards whenever they inspect a workplace which does not seem to manage the risks of passive smoking.

Rather than formal action via Improvement Notices, the officer is likely to deal with the matter informally through advice and guidance. They are recommended to encourage the development of smoking policies and to help and assist the employer in putting such a policy into practice.

Enforcement officers can require Risk Assessments to be completed and could serve an Improvement Notice requiring the completion of a Risk Assessment. This action is particularly likely if an employee has complained to the enforcing authority, e.g. Environmental Health Department.

If necessary, the officer could consult the Employment Medical Advisory Service who will be able to assist in identifying any acute health hazards from passive smoking in particular cases, e.g. inadequate ventilation may exacerbate the risk.

Enforcement action under the Health and Safety at Work Etc. Act 1974 could be considered if any employer fails to follow the advice and guidance given by the enforcing officer.

12

Fire safety

What are an employer's key responsibilities in respect of fire safety?

Fire is a hazard in the workplace and, like any other hazard, the employer has a duty to protect employees from the risks.

In respect of the hazard, i.e. the likelihood or potential to cause harm, fire is a high and serious hazard as it has the potential to kill or severely injure many people.

But the risk of a fire breaking out is very much in the control of the employer and effective fire safety procedures will reduce the potential or risk of fire to acceptable levels.

Legislation exists to protect employees and others from the risk of fire and employers have duties to comply with the requirements.

The main legislation covering fire safety is detailed below.

The Fire Precautions Act 1971

This Act is the principal piece of legislation to govern fire safety throughout all premises and places duties on owners and occupiers of buildings.

The principal requirement of the Act is to require the Fire Authority to issue Fire Certificates to all premises which meet the criteria.

The Act also includes provisions for the Fire Authority to carry out routine inspections, take legal action for contraventions, serve statutory notices, close down premises and prosecute owners or occupiers.

Fines for contravention of the Act can be unlimited.

The Fire Precautions (Workplace) Regulations 1997 (amended in 1999)

These Regulations require employers to specifically consider the risks of fire to their employees.

The Regulations require employers to:

- carry out fire Risk Assessments of the workplace
- prepare emergency procedures
- maintain fire-fighting equipment
- carry out employee training
- prepare an emergency plan
- co-operate with other employers
- establish reliable methods of calling the emergency services.

The Fire Precautions (Hotels and Boarding Houses) Order 1972

This Order was made under the provisions of the Fire Precautions Act 1971 and requires all premises which are used as hotels or boarding houses where sleeping accommodation is provided for more than six people (whether guests or staff) to have a Fire Certificate.

If accommodation is provided elsewhere than on the ground or first floors, for any number of guests or staff, a Fire Certificate will also be needed.

When do premises need a fire certificate?

The Fire Precautions Act 1971 requires certain premises to be issued with a Fire Certificate from the appropriate Fire Authority.

If the premises meet the following criteria, a Fire Certificate must be in existence. If one has not been granted but has been applied for, then the law is being met, provided suitable and sufficient fire safety measures are being practiced within the premises. An offence is committed if a Fire Certificate is legally required but has not been applied for.

Fire Certificates are required for all premises where:

(a) twenty or more persons are employed at any one time

or

(b) ten or more persons work at any one time in the building elsewhere than on the ground floor.

If premises have a Fire Certificate, does the employer have to complete a fire Risk Assessment?

Yes. A fire Risk Assessment is required for all premises in which people are employed to work unless they are covered by special legislation or are exempt under the Regulations.

Premises which are exempt from the need to have fire Risk Assessments are:

- workplaces only used by the self-employed
- domestic premises
- construction sites
- any aircraft, locomotive or rolling stock in use as a means of transport
- mines other than surface buildings
- agricultural or forestry land
- offshore installations
- ships within the meaning of the Merchant Shipping Act 1995.

What will a Fire Certificate cover?

- The use of the premises covered by the Fire Certificate.
- The means of escape in case of fire.
- The means to ensure that the means of escape can be safely and effectively used at all material times, including measures to restrict the spread of fire, smoke and fumes, escape lighting and directional signs.
- The type, number and location of fire-fighting equipment.
- The type, number and location of fire alarms.

At the discretion of the Fire Authority, the Fire Certificate may incorporate the following requirements:

- means of escape to be properly maintained and kept free from obstruction
- that all fire precautions, e.g. fire-fighting equipment, are properly maintained
- all employees to be given appropriate training in what to do in the case of a fire, and that records are kept of that training
- the number of people who may be in the premises at any one time will not exceed a specified number — the permitted number
- any other requirements thought necessary in order to protect persons from the risk of fire in the premises.

As an employer, am I responsible for the safety of members of the public in respect of fire hazards?

The Fire Precautions (Workplace) Regulations 1997 (as amended) require an employer to consider the hazards and risks to their *employees* from fire while they are at work. There is no duty under these Regulations to consider the public.

However, the general duty for Risk Assessments under the Management of Health and Safety at Work Regulations 1999 requires employers to consider the hazards of their work activity or undertaking to *other* people, including the public.

The amended version of the Fire Precautions (Workplace) Regulations requires a fire Risk Assessment to be carried out under the principles of general risk assessment contained in the Management Regulations.

Obviously, it is sensible to incorporate the hazards of fire to all persons resorting to premises or affected by the business undertaking.

How should I carry out a fire Risk Assessment?

Fire Risk Assessments should be approached in the same way as all Risk Assessments and the following steps can act as a guide.

Risk assessment is a five-step process:

Step 1: identify the hazards
Step 2: identify the people at risk
Step 3: remove or reduce the risk
Step 4: assess fire safety arrangements
Step 5: monitor and review.

The following checklists are not prescriptive — employers must use their judgement in deciding which actions, systems or procedures best qualify as being necessary, adequate or sufficient.

Step 1: Identify the hazards

- Highly flammable liquids, cleaning materials, etc.
- Combustible waste.
- Combustible materials.
- Substantial areas of walls or ceilings covered with flammable linings, furniture, etc.
- Any other readily combustible materials in the premises, presenting significant hazard.

Do your premises contain any of the following sources of heat?

- Work activities, including cooking, welding, flame cutting, frictional heat (including maintenance/building work).
- Use of oil/gas burning equipment.
- Presence of hot ducts or flues.
- Light bulbs and fittings near flammable materials.
- Electrical equipment with faulty or damaged wiring.
- Use of electrical extension leads.
- Use of portable heaters.
- Use of multipoint adaptors and electric sockets.
- The possibilities of arson (a significant cause of fire each year).
- Presence of any other sources of heat on the premises.

Step 2: Identify the people at risk

Consider the following:

- Do you have staff working in remote areas of the premises or in areas of high fire risk?
- Are your staff training and arrangements for safe evacuation adequate?
- If premises are used for sleeping accommodation, have you provided:
  - adequate fire warning arrangements
  - short distances of travel to safety
  - protected escape routes
  - sufficient staff to assist in evacuation?
- Have you provided suitable notices telling staff what to do in the event of a fire?
- Where people, i.e. customers, have hearing, sight or mobility impairment, have you provided:
  - adequate numbers of staff to assist in an emergency
  - arrangements to ensure staff are given early warning of fire

  - short distances of travel to safety
  - exit routes of adequate width?
- If the premises are occupied by large numbers of people, in particular members of the public, have you provided:
  - adequate signposting and escape routes
  - sufficiently trained persons to assist in evacuation
  - appropriate communication, e.g. alarm bells, etc.?

Step 3: Remove or reduce the risk

Consider any of the following as options for removing or reducing risk, including:

- removal, reduction or substitution of combustible and highly flammable materials
- safe storage of combustible and highly flammable materials, including:
  - a fire-secure environment with fire detection and warning systems if appropriate
  - fire-resistant cabinets to protect valuable documents and also to stop spread of fire
- removal and treatment of flammable walls or ceiling linings
- replacement or repair of damaged, upholstered furniture
- housekeeping improvements, e.g. keep workplace free of rubbish.

You should also consider the following:

- replace naked flame radiant heaters with safer forms of heating
- separate sources of heat by fire-resisting enclosures
- check electrical safety arrangements:
  - is equipment suitable for the purpose?
  - is equipment regularly inspected and maintained?
  - are inspections formally recorded?
  - removal and reporting of damaged items

- set up a Permit to Work system for 'Hot Works' and control contractors
- install safe forms of lighting
- provide no-smoking areas or buildings
- place controls on sources of ignition
- review security arrangements.

Step 4: Assessment of fire safety arrangements

You should consider the following.

- Fire detection systems
  - make sure they work
  - if they have to be isolated for maintenance/building work, ensure that alternative arrangements are made.
- Formal fire safety planning, written instructions and effective communication.
- Operating procedures: do they reinforce good health and safety practices or conflict with them?
- How far would you need to travel to get to safety?
- Fire escape routes: protect them, never lock exits, consider if lighting would be adequate.
- Do you need to set up additional escape route and exits or reduce the number of people in an area of the premises?
- Do you need to provide additional exit and directional signs?
- Do you need to install additional emergency escape lighting?
- Do you need to make special provision for anyone with disabilities, or where hotel accommodation is provided?
- Do you have enough adequately trained staff to help in an evacuation?
- Have you provided enough information, e.g. fire action notices, details of assembly points, etc.?
- Are staff fully trained to use fire-fighting equipment and other suitable fire suppression systems (or, according to company policy, are staff aware that they should not attempt to use fire-fighting equipment)?

- Do you need to appoint fire wardens?
- Do you keep extinguishers regularly maintained, free from obstructions and regularly tested?
- Are your fire doors adequate: these should be able to hold fire back for a minimum of 30 minutes, be in the correct locations and should never be blocked or wedged open?
- Check fire alarms: what are the distances between fire alarm call points and how would your staff be alerted to fire, is there a sufficient number of alarms, can they be easily seen, make sure they are never blocked or obscured.
- Do you need to install automatic fire detection?

Step 5: Monitor and review

Consider the following.

- Do you carry out regular, thorough inspections?
- Do you review your assessments in the light of changing circumstances, for example:
  - new procedures or methods
  - new equipment
  - change of premises or change of use of premises
  - light or temporary building work or maintenance
  - major building work
  - new members joining the team?
- Have you reviewed your assessments following any health and safety incidents, no matter how minor?
- Finally, do you regularly review assessments as a matter of course?

What training do I need to give my employees in respect of fire safety?

Employers are responsible for ensuring that all employees receive suitable and sufficient training in fire safety. This requirement exists whether the premises have a Fire Certificate or not.

Information, instruction and training should be straightforward and non-confusing.

Records of all employee training should be kept.

Training must include the following:

- the action to take on discovering a fire
- how to raise the alarm and what happens next
- the action to take on hearing the fire alarm
- the procedures for alerting customers and visitors, including, where necessary, directing them to exits
- the arrangements for calling the fire brigade
- evacuation procedures for everyone to reach the assembly point at a safe place
- the assembly point
- the location and, where appropriate, the use of fire-fighting equipment
- the location of the escape routes, especially those not in regular use
- how to open all escape doors including the use of any emergency fastenings, e.g. break-glass bolts, etc.
- the importance of keeping fire doors closed to prevent the spread of fire, heat and smoke
- how to isolate machines, equipment, etc. as necessary
- how to isolate power supply sources
- why not to use lifts (if provided)
- evacuation of disabled customers or staff
- importance of general fire safety and good housekeeping
- understanding the fire Risk Assessment, where it is kept, when it needs to be updated, etc.
- the control measures in place in the premises

- the need to report equipment faults and malfunctions
- how to deal with spillages.

Training should be given to employees on induction and then at least annually so that they are familiar with the requirements. Training also needs to be undertaken when any part of the premises changes, e.g. extensions are built, or when the service offered is changed, e.g. the functions or the work practices change.

What are some common sense, practical steps to implement regarding managing fire safety in premises?

Poor practices will often lead to a fire occurring and while the precautions for the premises may prevent the rapid spread of the fire and may allow people to escape unharmed, the practices of staff could make the difference between life and death.

Good management reduces the risk and suggested good management practices are given below.

Maintenance of plant and equipment

Keep all equipment well maintained. Develop a planned maintenance scheme.

Keep ventilation grills clean and free of dust build-up. Also keep clear of obstructions.

Keep all parts in good condition so as to avoid frictional heat build-up from poorly moving parts.

Keep electrical connections in good condition and make visual checks frequently.

Keep earth bonding leads in place and connected between equipment as advised by an electrician so as to avoid static electricity.

Storage and use of flammable materials

Check labels of all cleaning fluids, etc. to see whether they display the 'flammable' symbol.

Keep all such materials away from any heat source, i.e. do not store them next to cookers or other heat-generating equipment.

Do not store materials in the boiler and plant room.

Consider the fact that aerosol droplets, etc. can travel in the air and a source of ignition could be some way away from the actual storage or use area.

Do not use flammable liquids or substances near open flames, on electrical appliances, etc.

If flammable liquids are to be stored, keep them in a metal or fireproof container away from other materials, or keep them on a separate shelf away from other chemicals.

Never keep anything flammable anywhere near a fire escape route, fire door, etc.

Keep containers covered or keep caps on bottles, etc. — remember fumes and vapours can travel.

Do not store LPG or anything similar inside the premises unless it is connected to the heating appliance and is in use.

Store all LPGs in designated, locked and secure outside storage areas, preferably well vented to the external air.

Repairs, refurbishments, works, etc. and those involving heat-generating equipment

Operate a Permit to Work or Hot Works Permit procedure.

Check the area for flammable materials beforehand.

Have fire extinguishers readily to hand.

Supervise the works with a competent person.

Check one hour after works have finished that all is well.

Do not disconnect fire detection equipment unless absolutely necessary, but if so, reconnect it at the end of the works — as defined in the Hot Permits procedure.

Ensure all employees, contractors, etc. are familiar with the fire emergency procedures, fire exit routes, fire-fighting equipment locations, etc.

Remove rubbish and refuse regularly from the work site.

Ensure fire exit routes and doors remain unobstructed.

Keep fire doors shut.

Electrical equipment

Do not overload plugs, sockets and extension leads. Follow the maxim — one socket, one plug. Do not use multi-plug adaptors — use a properly fused 3 or 4 gang trailing socket and lead.

Visually check all appliance leads, cables, plugs and sockets regularly prior to use. Take out of use any suspect equipment and commission repairs or replacements. Seek guidance from either the company's maintenance department, electrician or maintenance contractor. Do not take risks for not only could there be a fire, but electric shock and electrocution are most likely from faulty equipment.

Check equipment is fitted with the correctly rated fuse.

Ensure light bulbs are the correct wattage for the lamp shades.

Do not use electrical equipment in areas where the atmosphere could be hazardous, e.g. after using flammable aerosols, etc. Ventilate the area first.

Do not store flammable materials near to electrical appliances which give off heat or which could build up static electricity.

Smoking materials

Keep a metal bin for emptying ash trays. Do not dispose of ash tray contents into ordinary waste bins, kitchen waste sacks, etc.

Have a fire extinguisher handy so that if a fire does start in a waste bin, it can quickly be controlled.

Discourage customers from discarding cigarettes other than in ash trays.

Introduce a smoking policy.

Waste and rubbish

Flammable waste and refuse must not be stored, even as a temporary measure, in escape routes such as corridors, stairways or lobbies, or where it can come into contact with potential sources of heat.

Accumulations of flammable refuse and waste should be removed regularly throughout the day and, in particular, from the kitchen where the risk of fire is greatest. Remove waste to external storage areas.

Ensure that all waste is kept in containers with lids so that accidental ignition sources can be avoided.

Keep waste away from garden seating areas. Ensure that there are facilities for customers to extinguish cigarettes, etc. within garden areas as this will prevent them using available waste bins, etc.

Assembly points

All premises used for employment and where members of the public are admitted must have an identified safe place where people can be directed to in an emergency and where numbers can be accounted for. This is usually known as an 'assembly point' or sometimes as the 'muster area'.

Make sure such a place is identified as part of your fire safety plan. The assembly point must be outside the building and far enough away so that people are not put at risk from the fire. All staff must know the assembly point and must report to the responsible person for a roll call. This will ensure that you know if anyone is missing, e.g. they could have been in the cellar and not heard the alarm (if this is the case, you will need to make sure that improvements in the fire warning system are implemented).

If the premises have a Fire Certificate, the assembly point is identified by the Fire Officer and is included in the schedule attached to the Certificate.

Suitable 'safe area' safety notices should be displayed in green and white.

What are the main requirements for means of escape in case of fire?

The premises must have alternative means of escape whereby people can leave the building other than by the route by which they came in.

Means of escape must be suitable for the numbers of persons who will need to escape from that area of the premises and, generally, must be at least one metre wide.

Doors must be openable in the direction of travel wherever possible, i.e. outwards, and must be easily opened at all times. Fire exit doors must *never* be locked, padlocked, jammed shut or obstructed.

A panic bolt/bar, quick release catch or thumb turn must be fitted. If doors need to be locked for security reasons, a break-glass bolt is required.

Fire exit doors must be suitable for the purpose, be fire-resisting for at least 30 minutes, be close fitting and fitted with intumescent strips (these expand in the heat and stop smoke seeping through the door frame).

Suitable signage must be displayed, i.e. a pictogram only or a pictogram and words indicating a fire exit door, displayed in green and white.

A 'means of escape' is the whole route from the area being escaped to the safe area outside the building. Corridors, passageways, etc. must be kept clear and unobstructed. They must be maintained in good repair, have no trip hazards, etc. Where there is a change of level, this must be clearly visible.

All means of escape must be clearly and adequately lit so that escapees can see where they are going even when the electric lights have gone off. Emergency lighting is a requirement in certain circumstances and reference will be made in the Fire Certificate.

All means of escape must open out into safe areas, i.e. means of escape cannot take you back into the building nor into a closed space without an exit.

Staff areas will need a means of escape and the rules about ease of use and unobstructed routes apply equally to staff areas as they do to customer areas.

Means of escape from plant rooms and roof plant areas also need to be assessed to ensure that egress can be made from the areas quickly and safely.

What fire signage needs to be displayed in premises to which fire safety legislation applies?

All signs should be in a position where they are seen easily and clearly.

Signs must now legally contain a pictogram and may contain directional arrows. Words can be used if necessary with the pictogram but fire exit signs with words only are now illegal.

- Fire warning signs are green with white writing.
- Fire exit signs are also green with white writing.
- Fire door signs, etc. are blue with white writing.
- Assembly point signs are green with white writing.
- Fire action notices are blue with white writing.

A fire action notice must be displayed adjacent to any fire call point. This notice must advise people what to do in the event of a fire.

What is a Fire Safety Policy?

A Fire Safety Policy sets out what you will do to manage the risk of fire within the premises, what you will do in an emergency, how you will maintain fire-fighting equipment, means of escape in case of fire, etc.

The Fire Safety Policy can be part of the overall Health and Safety Policy and contained in that document, or it can be a standalone policy.

A Fire Safety Policy is particularly important if the premises are open for the purposes of public entertainment. In these cases, a Fire

Certificate will probably have been issued for the premises. The Fire Certificate will contain standards of fire safety to be adopted within the premises.

A Fire Safety Policy should contain the following:

- description of the business
- assessment of fire risk, including Risk Assessments
- identification of hazards
- names of fire wardens
- emergency procedures
- monitoring procedures
- locations and types of fire-fighting equipment
- control measures
- staff training.

What equipment needs to be provided for emergency fire-fighting?

The fire Risk Assessment should determine what emergency fire-fighting equipment is necessary in the premises and where it will best be located.

If the premise have a Fire Certificate, fire extinguisher locations and types will be indicated on the plan attached to the Certificate. Fire extinguishers must ideally be wall-mounted, easily visible and accessible.

A 'fire point' with an appropriate sign is recommended, i.e. a place where two or three fire extinguishers are located adjacent to a fire alarm point.

Changes to legislation have required all new fire extinguishers to be red, with a coloured identification panel of either blue, red, cream, green or black, dependent on the contents of the extinguisher.

Existing fire extinguishers in red, cream, blue, black or green (quite rare these days) are still legal until such time as they need to be replaced.

Fire extinguishers must be serviced annually and need to be refilled.

If a fire extinguisher is used for any reason, it must be discarded until refilled. Never use or put back a half-filled or discharged fire extinguisher.

Fire extinguisher types

- Water — suitable for combustible materials such as paper, wood, textiles.
- Powder — suitable for all types of fire, including fires in electrical appliances.
- Foam — suitable for liquid fires.
- Carbon dioxide — suitable for electrical fires.
- Fire blanket — suitable for fat fryer fires, oils and for burning clothing.

13

Safe systems of work

What is a safe system of work?

There is no legal definition of what constitutes a safe system of work and it will be a matter of 'fact and degree' for the Court to determine.

Precedence was, however, set in the Court of Appeal in the 1940s when the then Master of the Rolls said:

> I do not venture to suggest a definition of what is meant by system. But it includes, or may include according to circumstances, such matters as physical lay-out of the job, the sequence in which work is to be carried out, the provision … of warnings and notices and the issue of special instructions.
>
> A system may be adequate for the whole course of the job or it may have to be modified or improved to meet circumstances which arise: such modifications or improvements appear to me to equally fall under the heading of system.
>
> The safety of a system must be considered in relation to the particular circumstances of each particular job.

This means that a system of work must be tailored for each individual job.

What is the legal requirement for safe systems of work?

The Health and Safety at Work Etc. Act 1974 sets out specifically in Section 2 that the employer is responsible for:

> the provision and maintenance of plant and systems of work that are, so far as is reasonably practicable, safe and without risks to health.

The Confined Spaces Regulations 1997 also require employers to establish a safe system of work if work and entry into confined spaces cannot be avoided.

What are the provisions for a safe system of work?

Generally, developing a safe system of work will involve:

- carrying out a Risk Assessment
- identifying hazards and the steps which can be taken to eliminate them
- designing procedures and sequences which need to be taken to reduce exposure to the hazard
- considering whether certain things or actions need to be completed before others
- designing Permit to Work or Permit to Enter systems
- writing down the procedure
- training employees and others.

Is a Method Statement the same as a safe system of work?

Generally, the two are similar and a Method Statement is a written sequence of work which should be followed by the operator in order to complete the task safely.

This is the same as a 'safe system of work' which is a sequence of events needed in order to reduce or eliminate the risks from a hazard which, in itself, cannot be eliminated.

Method Statements are common in the construction and maintenance industries and they are often required under construction laws, e.g. demolition works must always be accompanied by Method Statements.

The Control of Asbestos at Work Regulations 2002 require all work with asbestos to be supported by a work plan or plan of work, i.e. Method Statement or safe system of work.

When employees have to undertake hazardous tasks or when they have to work in hazardous environments, it is incumbent on the employer to ensure the safety of their employees and others. They must therefore decide *how* the job is to be done in order to ensure that their employees are kept safe.

What is a Permit to Work system?

A Permit to Work, or Permit to Enter, is a formal system of checks which records that the safe system of work which has been developed for the process is implemented.

The 'Permit' process usually applies to hazardous areas and is commonly used for:

- entering and working in confined spaces
- working of electrical plant
- working on railways or traffic routes
- working in chemical plants
- working in hazardous environments
- Hot Works.

It is a means of communication between site management, supervisors and those carrying out the hazardous work.

Essential sections of a Permit to Work are:

- clear identification of who may authorise particular jobs
- limitations in respect of anyone's authority
- clear guidance as to who is responsible for determining the safety procedures to be followed
- clear guidance as to what safety precautions are necessary
- details of emergency procedures
- information which must be relayed to site operatives
- instructions, training and competency requirements must be specified
- the condition for which the Permit is relevant
- the duration of the Permit
- the 'hand back' procedure
- monitoring and review procedures.

There are no set forms to use — employers should devise their own but the HSE have a free template associated with the ACOP and Guidance on Working in Confined Spaces.

Case study

Three men employed by a small plumbing and drainage company were called to unblock a sewer. The men lifted the manhole cover to find that the drains were about 3 m down. The step irons looked alright and the first operative, the 17-year old nephew of the company owner, descended into the sewer to unblock it with his rods.

When he got to the bottom of the shaft he collapsed.

His two colleagues panicked and descended into the sewer as well as rescue him, fearing that he might drown. Both men were overcome but managed to shout before passing out.

Fortunately, the Site Foreman had arrived and realised something serious was wrong. He called the emergency services.

The 17-year old operative was dead by the time he was rescued. The other two operatives died two days later in hospital.

They had died of hydrogen sulphide gas poisoning — a deadly poisonous gas which, the stronger and more lethal it is, the more odourless it becomes.

The company owner was prosecuted for health and safety offences — for failing to have a safe system of work to protect his employees while they were at work.

At the very least, the HSE said, there should have been a safe system of work operated by a Permit to Enter/Work system.

Gas monitoring/detection equipment should have been used, the sewer sludge should have been agitated to release any build-up of toxic gas, there should have been emergency procedures, breathing apparatus, training and so much more.

14

Risk Assessments

Is it a legal requirement to carry out Risk Assessments?

Yes. Every employer (and the self-employed) has a duty to undertake an assessment of the hazards and risks associated with work activities and to implement controls which either eliminate the hazard or reduce the risk to acceptable levels.

The Management of Health and Safety at Work Regulations 1999 set down the key requirements for Risk Assessments, but Risk Assessments are also required under:

- Control of Substances Hazardous to Health Regulations 2002
- Control of Lead at Work Regulations 2002
- Control of Asbestos at Work Regulations 2002
- Dangerous Substances and Explosive Atmospheres Regulations 2002
- Noise at Work Regulations 1989
- Health and Safety (Display Screen Equipment) Regulations 1992
- Manual Handling Operations Regulations 1992
- Fire Precautions (Workplace) Regulations 1997/99
- Personal Protective Equipment at Work Regulations 1992.

The provision for Risk Assessments under the Management Regulations is general but completing a Risk Assessment under these Regulations may satisfy the requirements of the specific regulations. Conversely, completing the specific assessments will broadly satisfy your duties under the Management Regulations.

Does absolutely every single job activity require a Risk Assessment?

No, although it sometimes feels like that!

The requirement of the Management of Health and Safety at Work Regulations 1999 is as follows:

> Every employer shall make a suitable and sufficient assessment of:
> (a) The risks to health and safety of his employees to which they are exposed whilst they are at work, and
> (b) the risks to the health and safety of persons not in his employment arising out of or in connection with the conduct by him of his undertaking,
>
> for the purpose of identifying the measures he needs to take to comply with the requirements and/or prohibitions imposed upon him under the relevant statutory provisions and by Part II of the Fire Precautions (Workplace) Regulations 1997 (amended).

If a work activity does not pose any health and safety risks then there is no need to carry out a Risk Assessment, although a Risk Assessment of sorts will be carried out in order to establish that the job task has no hazards and risk attached to it.

Increasingly, however, it is considered best practice to undertake Risk Assessments for all job tasks because, even though the statutory laws may not require them, the need to provide a duty of care under civil law makes Risk Assessments a valuable defence tool.

Is there a standard format for a Risk Assessment?

No, mainly because Risk Assessments are individual to job tasks and need to be 'site-specific'!

The HSE do publish guidance on how to complete Risk Assessments and they include a Risk Assessment template.

There is no right or wrong way to complete a Risk Assessment. The law requires that it is 'suitable and sufficient'.

A Risk Assessment must contain suitable information to be useful to an employee to understand what hazards they may be exposed to when carrying out the task.

Generally, any format which includes the following will be suitable:

- a description of the job task
- location of activity
- who will carry it out
- who else might be affected by the task
- what are the hazards identified
- what could go wrong
- what might the injuries be and how severe might they be
- how likely are the risks
- what can be done to reduce or eliminate the hazards
- what information do employees or others need to work safely
- when might the Risk Assessment be reviewed.

What do the terms 'hazard' and 'risk' mean?

A hazard is something with the potential to cause harm.

The risk is the likelihood that the potential harm from the hazard will be realised.

The extent of the risk will depend on:

- the likelihood of the harm occurring
- the potential severity of that harm (resultant injury or adverse health effect)

- the extent of people who might be affected, e.g. several people, vast groups, communities at large (e.g. from chemical releases).

What does 'suitable and sufficient' mean?

The phrase suitable and sufficient is not defined in the Regulations nor within the Health and Safety at Work Etc. Act 1974.

The Approved Code of Practice on the Management of Health and Safety at Work Regulations 1999 states that:

> The level of risk arising from the work activity should determine the degree of sophistication of the Risk Assessment.

Insignificant risks can generally be ignored, as can routine activities associated with life in general.

Risk Assessments are expected to be proportionate to the hazards and risks identified.

Enforcement Officers do not expect to see huge volumes of paperwork — the simpler the Risk Assessment and the clearer the information, the easier the employee will find following safe procedures.

When should Risk Assessments be reviewed?

The Management of Health and Safety at Work Regulations 1999 requires a Risk Assessment to be reviewed if:

- there is reason to suggest that it is no longer relevant or valid
- there has been significant change in the matters to which it relates.

If changes to the Risk Assessment are required, the employer has a duty to make the changes and reissue the Risk Assessment.

Employers are not expected to anticipate risks that are not foreseeable.

If events happen, however, which alter information available or the perception of risk, the employer will be expected to respond to the new information and assess the hazards and risks in the light of the increased knowledge.

Accidents and near misses should be investigated as these incidents will indicate whether more knowledge is available on the hazard or risk associated with the job. The Risk Assessment may need to be reviewed because:

- something previously unforeseen has occurred
- the risk of something happening or the consequences of the event may be greater than expected or anticipated
- precautions prove less effective than anticipated.

New equipment, new working environment, new materials, different systems of work, etc. will all require existing Risk Assessments to be reviewed.

Who can carry out Risk Assessments?

Whoever carries out a Risk Assessment must be 'competent' to do the job.

Competency is not defined in the legislation but such persons must have suitable training and experience or knowledge in order to identify hazards and risks associated with the job tasks.

Risk Assessments can be undertaken by managers, employees, consultants, individuals or teams.

Complex work processes and tasks may need external specialist consultants to assist with the Risk Assessment. But the *employer* is always responsible for ensuring that a suitable and sufficient Risk Assessment is completed.

Risk Assessments could be divided into different work or departmental areas so that those who know what happens in their department use their expertise to complete their own assessments.

How should a Risk Assessment be completed?

A Risk Assessment is really a logical review of what is done by whom, what can go wrong, what the consequences are, what can be done better, what procedures can be put in place to reduce the hazards and risks.

The HSE advocate a *five-step* approach:

Step 1: Identify the hazards
Step 2: Determine who could be harmed and how
Step 3: Evaluate the risks and decide if control measures are needed — if they are already in place are they adequate or does more need to be done
Step 4: Records your findings
Step 5: Review and revise the assessments as appropriate.

What needs to be done to identify hazards?

Hazard identification is not a complicated process and should not become so.

Walk around the workplace and look at what could cause harm to both employees and others. Also, consider whether there would be any additional risks to young people and pregnant women or nursing mothers.

Concentrate on significant hazards which have the potential for quite serious injury or ill-health or which could affect many people.

Speak to employees and ask them what they would identify as hazards associated with their jobs — run a hazard identification campaign.

Common hazards

- Fire
- Slips, trips and falls from uneven or slippery floor surfaces
- Falling objects
- Electricity
- Hazardous substances
- Transport vehicles
- Using equipment
- Excessive temperatures
- Lack of ventilation
- Poor lighting
- Confined spaces
- Explosive atmospheres
- Changes in floor level
- Use of portable equipment
- Use of hand tools

Hazards could be associated with:

- equipment — how it is used, guards, controls, noise
- work processes — how things are done, systems to be followed
- environmental conditions — floors, heating, ventilation, etc.
- materials in use — chemicals, gases, substances, etc.

What people must be considered as being exposed to the risks from work activities?

- Employees of the employer
- Young workers and those on work experience
- New and expectant mothers
- Cleaners — whether contract or in-house
- Visitors to the premises
- Maintenance workers — both contract and in-house
- Members of the public
- Employees of other employers with whom you share the building or premises
- Delivery drivers
- Sales representatives
- People with disabilities —extra control measures may be needed to protect them from risk
- Peripatetic workers — those working away from the office, usually visiting other workplaces or people's homes, e.g. midwives
- Volunteers.

Must Risk Assessments be categorised into high, medium or low risks?

Not necessarily by law, but it is good practice to identify the extent of harm that an employee or other person could be exposed to.

Even after all precautions have been taken, some risk, i.e. potential cause of injury or ill-health may remain. This is often referred to as:

residual risk.

Residual risk is either 'high, medium or low', or 'very likely, probable or unlikely'.

Some risk management approaches allocate numerical scores to various types of risk and the severity of those risks. By multiplying one score by the other they arrive at a 'risk rating'. Scores above a set target become unacceptable and measures must be put in place to reduce the risks.

What are some of the common control measures which can be put in place to reduce the risks from job activities?

The aim of risk assessment is to reduce the residual risk associated with a task to as small as possible.

First, try to eliminate the hazard altogether — why do something or use something if you don't have to?

Where the hazard cannot be eliminated it must be reduced to acceptable levels by implementing *control measures*.

A common approach is by following the 'hierarchy of risk control', namely:

- try a less risky option, i.e. substitute something less hazardous
- prevent access to the hazard, e.g. by guarding
- organise work to prevent exposure to the hazard
- issue personal protective equipment
- protect the workforce as a whole, e.g. through exhaust ventilation

- provide welfare facilities to aid removal of contamination, to take rest breaks, etc.
- provide first aid facilities.

What will an Enforcement Officer expect from my Risk Assessments?

Enforcement Officers will want to see that you have:

- completed Risk Assessments
- considered site-specific issues
- completed comprehensive checks of the workplace
- involved workers
- considered the hazards and risks to others
- dealt with immediate hazards to reduce risks
- a system for reviewing Risk Assessments
- suitable records
- provided information, instruction and training to your employees
- introduced suitable and sufficient measures.

Site-specific Risk Assessments are probably the most important aspect. EHOs and HSE Inspectors are not keen on 'generic' Risk Assessments unless steps have been taken to ensure that any special site hazards and precautions have been added to the Risk Assessment.

To Enforcement Officers, risk assessment is not a paper exercise to be pulled off the shelf in from a manual. It is a pro-active approach by an employer to consider what could harm his employees and others and what measures he intends to take to reduce the risks of injury and ill-health.

15

Managing contractors

Why are employers responsible for managing contractors?

Employers have duties under the Health and Safety at Work Etc. Act 1974 for the health and safety of persons who are not in their employ but who may be affected by their undertaking.

There will often be a shared responsibility between employers for the safety of employees, and the self-employed, and for persons who use or resort to premises.

The principle that organisations (employers) retain responsibility for the safety of contractors working on their premises was established in the Associates Octel case, heard in the House of Lords in 1996.

The case involved a maintenance contract in respect of some tanks, which were classified as a confined space. Associate Octel's employee was injured because he used the wrong equipment in a hazardous environment.

The factory plant itself was closed for the annual summer shut-down and the maintenance contractor was the only one working in the area.

He was cleaning the inside of the tank with acetone and was using an electric light by which to see. Having nothing suitable to keep his acetone in, he retrieved an old bucket from the skip. The open container allowed the acetone to give off large quantities of flammable fumes. The environment was confined so the fumes could not disperse easily.

The light bulb broke. There was a flash fire as the flammable fumes and vapours caught fire. The maintenance engineer was badly burned.

Octel was prosecuted under Section 3 of the Health and Safety at Work Etc. Act 1974 for failing to ensure the safety of persons not in their employ.

Octel defended itself and said that the way that the maintenance contractor carried out the task was up to them as they had the duty under Section 2 of the Health and Safety and Work Etc. Act to ensure the safety of their own employees, and that Octel had no right to control or stipulate how they did it.

The case finally went to the House of Lords on appeal by Octel. Their appeal was rejected and they were found liable for the safety of the contractors.

Octel, in effect, employed the contractor because they were regular workers on the site and Octel provided them with safety equipment and required them to follow a safe system of work, via a Permit to Work.

Clear case law exists that employers have quite extensive duties for the safety of contractors working on their premises, especially if the jobs being undertaken are an integral part of the employer's business.

What needs to be done to manage contractors?

The HSE have produced guidance for employers and recommend a *five-step* approach.

Five steps to managing contractors' health and safety

Step 1: Planning

- Define the job.
- Identify hazards.

- Assess risks.
- Eliminate and reduce risks.
- Specify health and safety conditions.
- Discuss with contractor.

Step 2: Choosing a contractor

- Check that the contractor is competent for the job (ask questions, get evidence).
- Discuss the job, the site and site rules.
- Obtain a safety method statement.
- Decide if sub-contracting is acceptable and will be safe.

Step 3: Managing contractors on site

- Ensure that contractors sign in and out.
- Name a site contact.
- Reinforce health and safety information and site rules.

Step 4: Keeping a check

- Assess how much contact with contractors is needed.
- Is the job going as planned?
- Is the contractor working safely and as agreed?
- Have there been any incidents?
- Have there been changes in personnel?
- Are any special arrangements required?

Step 5: Reviewing the work

- How effective was the planning?
- How did the contractor perform?
- Record the findings.

16

Electrical safety

What are the key responsibilities of an employer in respect of electrical safety?

Electricity is a hazard and, every year, over 1000 accidents at work are reported to the enforcing authorities involving electric shock or burns while employees and others are using electrical equipment at work.

The general provisions of the Health and Safety at Work Etc. Act 1974 apply to all electrical equipment, i.e. employers have to ensure that plant, etc. is maintained in a safe condition.

The Electricity at Work Regulations 1989 place more specific duties on employers and 'duty holders' and address such issues as:

- live working
- maintenance of electrical appliances
- safe systems of work
- design and construction of electrical systems
- portable appliances
- strength and capability of equipment
- siting of electrical systems
- insulation, protection and excess current protection
- working space, access and lighting
- competency of operatives
- training, information and instruction.

The Provision and Use of Work Equipment Regulations 1998 will also apply to all electrical equipment in use at work.

What are the hazards associated with electricity?

Electricity can kill if used in a cavalier manner. It can also cause severe personal injuries, major electrical buns, heart attacks, etc.

The main hazards are:

- contact with live parts of equipment or cables causing shock, burns or death — normal mains voltage at 230 volts AC can kill
- electrical faults which cause fire due to overheating or sparks igniting combustible materials
- fire or explosion where electricity could be the source of ignition in flammable or explosive environments.

What are the risks from electricity?

Electricity, if used correctly, presents a fairly low risk — although it has a high hazard rating, i.e. it has the potential to cause harm — the reality of that harm being realised is quite low in most instances because common sense controls are usually in place.

The likelihood of the hazard causing injury will depend on the circumstances:

- what is being done
- by whom
- where
- in what conditions
- with what equipment?

Switching on the kettle in the staff kitchen is in itself potentially hazardous because the kettle runs on electricity which has the potential to cause harm. But, if the kettle is maintained in good condition, the lead is not defective, the plug is the right sort with the correct fuse and the environment in which it is to be used is 'normal', the risk of being electrocuted or suffering an electric shock is extremely low.

However, use a defective kettle with a split lead and wrong plug and fuse and switch on with wet hands and the risk rating for an electric shock rises dramatically!

So, with electricity, the risk really is dependent on a number of other factors. The safer these other factors, the smaller the risk of suffering from electric shock or other hazard.

It is important to understand that certain factors increase the risk from electricity:

- using electricity and electrical appliances in wet conditions or damp atmospheres
- using equipment in flammable or explosive environments
- using equipment or electrical systems at times of gas leak
- working on live currents
- using extension leads, especially if connected in series
- having work carried out by non-qualified and incompetent persons
- using inferior cabling and electrical materials, i.e. those not kite-marked
- using equipment out of doors
- using equipment in cramped conditions with a lot of earthed metalwork, e.g. inside tanks, silos, etc.

Is it a legal requirement to have all portable appliances tested every year?

No. The law, namely the Electricity at Work Regulations 1989, requires that all electrical equipment is maintained in a safe condition. It does not require an elaborate and frequent system of testing.

What is portable electrical equipment?

Generally, any electrical equipment, tools or appliances which have a lead and plug and which could, if necessary, be moved from place to place.

Examples include:

- photocopiers, printers, fax machines, scanners, etc.
- computers/PCs, shredders, etc.
- kettles, vacuum cleaners, pressure washers, televisions, videos, OHPs, desk lamps, etc.
- portable tools, power tools, etc.
- fans, dehumidifiers, heaters, etc.

What does an employer have to do to demonstrate that he has 'maintained' portable electrical equipment?

The HSE, in its numerous guidance documents, recommends that most portable appliances will need to be *visually inspected* as the first stage of a maintenance regime.

Most accidents involving portable appliances occur because the equipment, or more importantly, the lead and plug, is defective. Such defects are usually visible to the user and, provided the user has some training on the safe use of portable appliances, the equipment should not be used.

So, everybody should be trained in simple 'user checks' of electrical equipment and everyone should be instructed to carry out a 'user check' before use.

Look for:

- cuts or abrasions to cables
- loose wires
- broken plugs
- burning smells

- scorching of plugs or equipment casing
- coloured wiring showing from the plug
- incorrect fuses
- extension leads plugged into extension leads.

Employers need to determine that they have a system of user checks in place — keep training records to show that everyone has been trained in what to do.

After 'user checks', the next level of maintenance is 'formal visual inspection'. This is when the employer uses a 'competent person' to visually inspect the equipment and record the results. The competent person could be an employee trained in what to look out for. A simple test appliance could be used, called a PAT kit — Portable Appliance Test kit. This will indicate whether the 'earth continuity' of the equipment is satisfactory, unless, of course, the equipment is 'double insulated' and does not have an earth lead.

Certain equipment will also need regular 'combined inspection and testing' as the final stage of maintaining electrical appliances. This is usually carried out by qualified electricians and is normally undertaken at anything between one- and five-year intervals depending on the Risk Assessment.

Is it necessary to keep records of any inspection and testing?

The law does not require the keeping of records but it is a sensible thing to do and would provide you with a defence if someone were to accuse you of having faulty equipment.

The Provision and Use of Work Equipment Regulations 1998 require employers to demonstrate that they have maintained equipment in a safe condition and this can really only be done with a suitable record system.

A record system will help you to monitor what needs to be done and when. Computerised record systems could be used to implement a 'bring forward' system whereby equipment which is scheduled either for 'formal visual inspection' or 'combined inspection and testing' is listed with due dates for action, etc.

Is all electrical equipment to be treated the same?

No. As stated earlier, you must carry out a Risk Assessment. If equipment is subject to user abuse or is used in hazardous environments it will need far more attention than, for example, a photocopier.

The Risk Assessment should include the frequency for inspection and testing and what sort is needed, e.g.:

- user checks — every time equipment is used
- formal visual checks — every six months by Maintenance Department
- combined inspection and testing — every two years by external electrical company.

The Regulations not only apply to electrical appliances, but also to 'electrical systems' which includes plugs, leads, cables, etc.

As an employer, am I responsible for electrical equipment brought into work by my employees?

If the equipment is used at work because it supplements what you have provided, e.g. extra portable heaters in winter, then it must be classed as 'work equipment' and it should be subject to the same checks and inspection as the company's own equipment.

Ideally, employees should be discouraged from bringing personal equipment into work.

How can I tell if electrical equipment is damaged?

Around 95% of electrical equipment faults can be identified by user checks and a visual inspection.

Identify the equipment, where it is used, how it is used.

Keep records as they will be useful if things go wrong or you need to demonstrate 'due diligence'.

Disconnect the equipment first!

Check the equipment, cable and plug.

Look for:

- damage, e.g. cuts, abrasion to the cable covering
- damage to the plug, e.g. the casing is cracked or the pins are bent or missing
- non-standard joints, including taped joints or joins in the cable
- the outer sheath not being gripped where it enters the plug or equipment — i.e. the coloured insulation of the internal wires is showing
- bare wires, i.e. brass showing
- equipment being used in unsuitable conditions, e.g. wet conditions, dusty workshops
- damage to outer cover of the equipment or obvious loose parts, screws, etc.
- overheating — burns, scorch marks to both plug or equipment casing or cable
- take the plug apart and check for the correct fuse and that no temporary 'repairs' have been carried out, e.g. a nail used instead of a fuse, or tin foil
- correct wiring to the correct terminals, i.e.:
  - blue to neutral
  - brown to live
  - green/yellow to earth
- terminal screws are tight
- no signs of internal overheating
- no bare wires showing.

What should I do if equipment is found to be faulty?

Take it out of use immediately and have it repaired by a competent electrician.

If the equipment cannot be moved or removed, then ensure that it is suitably labelled with a hazard warning sign to state:

> *Electrical fault. Unsafe to use.*

Where it is safe to do so, disconnect the power supply. Or, 'lock off' the power supply — place a Prohibition Notice on or near the main power switch or supply.

It will not be an offence to have defective equipment within the workplace. It *will* be an offence if employees or others use equipment which is known to be faulty and where you failed to take suitable steps to prevent persons being exposed to danger.

Case study

A lift engineering company and a hotel were both fined after a worker died as a result of an electric shock from a lift motor he was repairing.

The lift engineer was working in the lift motor room. He turned off the electrical supply to the cabinet he was working on but had not turned off a separate, unlabelled switch to a piece of equipment inside the cabinet. The engineer's knee came into contact with the equipment and he received a 230 v shock. His body was not found until the following morning because he was working alone and there was no safe system of work in place, operated by the hotel.

The lift company had failed to carry out a suitable and sufficient Risk Assessment, as had the hotel company.

The Council, who brought the prosecution, stated that a number of control measures could have been instigated such as rubber matting, improved lighting and better signage on emergency and electrical supplies.

Total fines were £80 000 + costs, £50 000 for the lift company, and £30 000 for the hotel company.

The family of the man who died could instigate civil proceedings against his employer.

17

Personal protective equipment

What are employers' duties in respect of the provision of personal protective equipment for employees?

The Personal Protective Equipment at Work Regulations 1992 require employers to make a formal assessment of the needs of their employees for personal protective equipment (PPE).

Employers must provide suitable and sufficient PPE to employees free of charge if the work Risk Assessment identifies the need for PPE as an effective control in reducing hazards and risks.

Employers must ensure that any PPE they choose and purchase must comply with the EC Directive for manufacturers — i.e. that the products have been independently certified and have a 'CE' marking.

Appropriate selection of PPE is an important employer duty, as is maintenance, cleaning and training in the use of PPE.

Suitable accommodation must be made available for storing PPE so that its condition does not deteriorate.

Records must be kept of PPE issued to employees, maintenance undertaken, any inspection, testing and examination results.

What is personal protective equipment?

PPE is defined in the Regulations as:

> all equipment (including clothing affording protection against the weather) which is intended to be worn or held by a person at work and which protects him against one or more risks to his health and safety.

Examples of PPE include:

- safety helmets
- safety footwear
- gloves or gauntlets
- goggles or visors
- hi-vi clothing
- safety harnesses
- personal alarms.

If employees' health and safety will be adversely affected by the weather, then waterproof, weatherproof or insulated clothing will be subject to the Regulations.

Ordinary staff uniforms which are not needed for health and safety protection, sports clothing, etc. are not covered by the Regulations.

Is personal protective equipment always an adequate health and safety safeguard?

No. Personal protective equipment is not always effective in protecting employees from the hazards and risks associated with their jobs. Its effectiveness can easily be compromised by:

- not being worn properly
- poor sizing
- unsuitable specification
- poor maintenance
- wrong equipment for the job.

Personal protective equipment must be considered as a *last resort* in the hierarchy of risk control, i.e. only if there are no other effective means of eliminating or reducing the risk, should PPE be issued to employees.

Often, employees find PPE uncomfortable or cumbersome and they refuse to wear it. Also, there may be a culture among peer groups to avoid wearing it for fear of looking 'sissy', e.g. many construction sites and construction workers are reluctant to wear and use PPE effectively.

PPE must be readily available to all employees if it has been identified on the Risk Assessment as a control measure and it must be available *free of charge*.

How should the suitability of personal protective equipment be assessed?

To allow the right type of PPE to be chosen, the different hazards in the workplace need to be considered carefully. This will enable an assessment to be made of which types of PPE are suitable to protect against the hazard and for the job to be done. Your supplier should be able to advise you on the different types of PPE available and their suitability for different tasks. It may be necessary in a few particularly difficult cases to obtain advice from different sources — and, of course, from the PPE manufacturer.

The following factors should be considered when assessing the suitability of PPE.

- Is it appropriate for the risks involved and the conditions at the place where exposure to the risks may occur? For example, eye protection designed for providing protection against agricultural pesticides will not offer adequate face protection for someone using an angle grinder to cut steel or stone.
- Does it prevent or adequately control the risks involved without increasing the overall level of risk?

Case study

What PPE do tree surgeons need?

- Safety helmet — at risk of branches falling on their heads.
- Eye protection — at risk of being poked in the eye by branches, twigs or having dust, grit and flying debris penetrate their eye.
- Ear defenders — will use chainsaws and the noise levels may be above the action levels.
- Gloves — will be subject to cuts and grazes due to branches, etc. Gloves should have padded backs especially if a chainsaw is to be used.
- Body protection, e.g. to legs and arms.
- Clothing — needs to be close fitting to prevent it being caught and entangled on branches.
- Safety boots — non-slip soles, steel toe caps and soles, weatherproof but flexible boots are essential.

- Can it be adjusted to fit the wearer correctly?
- Has the state of health of those who will be wearing it been taken into account?
- What are the needs of the job and the demands it places on the wearer? For example, the length of time the PPE needs to be worn, the physical effort required to do the job and the requirements for visibility and communication.
- If more than one item of PPE is being worn, are they compatible? For example, does the use of a particular type of respirator make it difficult to get eye protection to fit properly?

What needs to be covered in respect of information, instruction and training in relation to personal protective equipment?

An employer must provide an employee with suitable and sufficient information, instruction and training so as to enable the employee to know:

- the risks which PPE will avoid or minimise
- the purpose for which and the manner in which PPE is to be used
- any action which the employee may take to ensure that the PPE remains effective.

Employees must have formalised training on the use of PPE and it will be essential to keep records of who was trained on what and when.

If an employee has an accident or suffers ill-health because they were not wearing PPE or not using it properly, then, as an employer, you will need to demonstrate that they knew how to use it and why, otherwise you could be held responsible because you had failed to train them properly.

Employees need to know:

- what the PPE for each job is
- why they are required to use it
- how it will protect them
- how to wear it safely
- how to carry out pre-use checks
- how to report defects and to whom
- how often it is to be maintained
- whether it is subject to formal inspection, examination and testing
- who is competent to carry out tests, inspections and maintenance
- how to wear or fit it properly
- how to adjust it for comfortable wear
- conditions which might affect its effectiveness, e.g. hard hats go brittle when exposed to sunlight
- at what intervals it has to be replaced
- disciplinary procedures for failing to follow the 'rules'
- 'no hat, no kit, no job'.

Keep detailed records of the training and any tool-box talks given to employees. As always with records, you will only realise that you need them when you are subject to a criminal investigation or when you want to defend yourself against a civil claim.

What needs to be done in respect of maintenance of personal protective equipment?

Equipment needs to be well looked after and be properly accommodated when not in use, for example, stored in a dry, clean cupboard or, in the case of smaller items, in a box. It should be kept clean and in good repair — the manufacturer's guidelines should normally be followed. Simple maintenance can be carried out by the trained wearer, but more intricate repairs should only be done by specialist personnel. To avoid unnecessary loss of time, it is advisable that suitable replacement PPE should always be readily available.

Top tips

- Check that there are no other, better and more effective ways to control risks, e.g. engineering controls.
- Check that PPE is available to all those identified by the Risk Assessment as needing it.
- Check that it offers adequate protection.
- Check that those using it are adequately trained and have information about the PPE.
- Check that it is properly maintained and defect-free.
- Check that it is returned to storage after use.

18

Legionnaires' disease

What is legionnaires' disease?

Legionnaires' disease is a respiratory disease caused by the inhalation of the legionella bacterium, legionella pneumophilia. There are other strains of bacteria which also cause the disease or an illness similar, e.g. Pontiac fever.

The symptoms are:

- flu-like
- malaise
- general pains
- headache
- temperature — often up to 40°C
- dry cough
- nausea, vomiting, diarrhoea (at times)
- severe respiratory infection.

The onset of the symptoms varies and the common incubation period is between two and ten days.

Is everybody affected to the same extent?

No, not necessarily so. Certain groups of the population are more susceptible to the disease than others. Usually, healthy people will generally recover from the illness once antibiotic treatment has been given.

However, those most at risk are people with immune suppressed conditions, e.g. suffering from cancer, respiratory diseases, kidney disease, etc; those suffering from diabetes; people who smoke and those with chronic dependent illness, e.g. alcoholism.

The common age range for those most at risk seems to be between 40 and 70 years.

Where are legionella bacteria found?

Legionella bacteria are widespread in natural watercourses including rivers, streams and ponds. The bacteria can also be found in soil, although it is far more prevalent in water systems.

Commonly, legionella bacteria are found in hot and cold water systems, particularly those which store water and recirculate it, e.g. air conditioning systems, spa baths, etc.

How is the infection spread?

The legionella bacteria are spread by airborne droplets of water or water vapour, e.g. sprays, mists.

The water droplets in spray or mist form are inhaled by individuals and the infection takes hold within the body. The amount of bacteria which needs to be inhaled in order to develop the symptoms is not defined — it will vary from person to person depending on their general state of health.

Are outbreaks of legionnaires' disease common?

No, not especially so, although when they do occur they can affect large numbers of people.

For an outbreak of legionnaires' disease to occur, a sequence of events has to take place:

- conditions have to exist which suit the multiplication of the bacteria
- water temperatures have to be between 20°C and 45°C
- sludge, scale, rust, algae or other organic matter must be present to provide the bacteria with nutrients
- means of creating breathable droplets of water must exist
- contact with the infected droplets by a susceptible person must occur.

During 2002, a major outbreak of legionnaires' disease occurred in Barrow in Furness, Cumbria. Several people died from the disease and many were affected by the illness. The cause was traced to a polluted water treatment plant at the town's Leisure Centre complex. The outbreak affected a large number of people because the infected droplets of water were being discharged through a ventilation grill and everyone who walked past the centre or who were in the town centre over a period of days were at risk of infection.

What are an employer's legal responsibilities?

The general duties of the Health and Safety at Work Etc. Act 1974 apply, in particular, Section 2 — the duty to ensure a safe place of work.

In addition, the Management of Health and Safety at Work Regulations 1999 and the Control of Substances Hazardous to Health Regulations 2002 require employers to consider the hazards and risks to employees from work activities, including any exposure to biological organisms.

Employers must complete Risk Assessments for the likely exposure of employees and others, to legionella bacteria. If there is a risk of exposure, the employer must implement control measures. Approved Codes of Practice have been issued on the Prevention and Control of Legionellosis (including legionnaires' disease) as well as a Guidance on the Control of Legionellosis including Legionnaires' Disease.

Employers also have duties under the Notification of Cooling Towers and Evaporative Condensers Regulations 1992. These Regulations require that local authorities are notified of the location of wet cooling towers and evaporative condensers so that, if there is an outbreak, the Control Infection Team or Environmental Health Department know where to start their investigations for possible sources of contamination.

Also, if considered appropriate, the local authority can require employers or others to control or abate a 'statutory nuisance' if they believe a breach of the Environmental Protection Act 1990 has occurred. Any 'fumes or gases emitted from premises' and 'dust, steam, smell or other effluvia arising on industrial, trade or business premises' which is considered prejudicial to health or a nuisance can be the subject of enforcement action.

Employers therefore need to be mindful of environmental legislation as well as health and safety.

What do employers need to do to reduce the risk of a legionella outbreak?

The first step is to carry out a Risk Assessment to identify whether you have the hazard on the premises. All water supplies have the potential to be contaminated with legionella bacteria and so, mostly, the hazard cannot be eliminated at source.

Consider whether your employees and others are exposed to:

- hot water
- warm water — about 20°C or so

- showers
- air conditioning cooled by water
- humidifiers
- leisure facilities, e.g. spa baths, jacuzzis
- hot tubs
- water spraying operations.

Members of the public are equally at risk of exposure to legionella bacteria and if the business is involved in the following there is a greater risk:

- leisure services
- hotels
- care homes or health care
- residential homes.

Having established whether there is a water system present which could be contaminated with legionella bacteria, the next step is to establish the likelihood of exposure to the hazard, i.e. the risk.

Consider whether the water system:

- operates below 20°C
- operates above 45°C
- operates somewhere in the middle of those temperatures
- is the system regularly disinfected?
- is water present or generated in droplet, spray or mist form?
- is it regularly cleaned?

Remember, as with all bacteria, legionella need the opportunity to multiply to be infectious and if the conditions which are ideal for this are prevented, the risk will be significantly reduced.

Step three will be to establish that there is the potential risk of legionnaires' disease being spread and to determine the control measures which will eliminate the hazard or reduce it to acceptable levels.

The Risk Assessment must be in writing and regularly reviewed.

What are some of the control measures which can be put in place?

The following are a range of common sense control measures for all water systems which have the potential to be contaminated with legionella bacteria and spread legionnaires' disease. Suitable precautions will effectively reduce the risk of an outbreak occurring.

Hot and cold water services

- Keep water temperatures *below* 20°C or *above* 45°C.
- Fit thermostatic tap valves.
- Store hot water at 60°C or above.
- Circulate water at 50°C.
- Keep pipes well insulated and avoid cold water pipes being affected by hot pipes.
- Keep pipe runs short.
- Avoid 'dead legs'.
- Run water from all outlets regularly — flush the system though.

Cooling towers

- Notify the cooling tower to the local authority.
- Clean and disinfect the system at least every six months.
- Treat the water so as to prevent scale, algae and microbiological build-up.
- Take water samples for analysis.
- Follow manufacturers' instructions.
- Fit drift eliminators.
- Replace water-cooled systems with air systems.
- Use 'dip slides' to monitor microbiological activity weekly.

Other water systems

- Clean and disinfect equipment regularly.
- Descale shower heads — bacteria can survive on the scale.
- Keep temperatures as for hot and cold water.
- Disinfect water systems.
- Backwash and clean filtration systems weekly.

What happens if an outbreak occurs?

A full investigation is carried out by the local Control of Infection Team which usually comprises members from Environmental Health, The Health and Safety Executive, Public Health, Communicable Diseases Unit, Health Trust and, if fatalities have occurred, the police.

The need to identify the source of the infection is paramount — hence the requirement to notify water cooling towers.

Enforcement Officers have the powers to close buildings, prohibit the use of equipment or plant, etc. if they suspect that there is 'imminent risk' to people's safety.

Officers will visit premises and check records of maintenance, cleaning and bacteriological sampling. Temperature records of water are important.

Questionnaires will probably be sent to people who are known to have been affected or who have been in the vicinity of suspect buildings or in given areas.

Legionella bacteria can drift for considerable distances on the wind and so the source of the infection could be several hundred metres away from where people appeared to contract the disease.

Managers, employers, water service companies, etc, involved with the suspect plant or system could be interviewed under Section 20 of the Health and Safety at Work Etc. Act 1974, or under the Police and Criminal Evidence Act 1984, with a caution.

What should an employer do if bacteriological results come back positive for legionella?

Unless the employer is competent to interpret the results, they must consult and seek advice from experts so that the severity of risk can be determined.

The laboratory undertaking the analysis would be able to help with practical advice.

Generally, if legionella are identified it will indicate that something has gone wrong with the maintenance and cleaning regime. Shut down any systems so that you reduce the risk of the contamination spreading.

Drain down any water system, clean and disinfect all aspects of tanks, pipes, shower heads, taps, etc.

Seek advice from water treatment companies — they themselves have duties to act competently under the Approved Code of Practice.

19

New and expectant mothers and young workers

What are the legal requirements in respect of new and expectant mothers while they are at work?

The health and safety of new and expectant mothers is covered in the Management of Health and Safety at Work Regulations 1999.

Employers are required to assess the risks to health of work activities and to do what is reasonably practicable to control the risks identified. This includes any specific risks to certain categories of employees who may be more vulnerable to hazards because of a health condition, pregnancy or age.

If a hazard and risk cannot be controlled in any other way, an employer is required to make changes to the employee's working conditions, hours of work, etc. This will include offering suitable alternative work, or if that is not possible, the employer is required to suspend the employee (on relevant pay) for as long as is necessary to protect her health and safety and that of her unborn child.

If an employee normally works nights and has a certificate from a medical practitioner or midwife that working nights is not conducive to her health and safety, the employee must be suspended from the job for as long as medical opinion dictates is necessary. Where possible, suitable alternative day-time work should be offered before the night-time work is suspended.

Alternative work must be on the same terms and conditions as the woman's usual employment, even if the job is less skilled, etc.

The main legal requirement is to carry out a Risk Assessment of all the job activities carried out by the expectant employee with a view to assessing how, if at all, the employee's health and safety, or that of her unborn child, may be affected by the tasks she has to do.

What proof can I ask for to confirm that my employee is pregnant?

An employee who is, or alleges to be, pregnant must inform you in writing that they are so. Or, if they have given birth in the last six months they must inform you that they are a nursing mother.

As an employer, you may request in writing a certificate from a registered medical practitioner or registered midwife confirming the pregnancy. A reasonable time must be given for the employee to comply. If, within the reasonable time, no certificate is provided, as an employer you need do no more with regard to following specific health and safety requirements for pregnant employees until you have such a certificate.

What is the definition of a 'new or expectant' mother?

The phrase 'new or expectant employee' means an employee who is pregnant or who has given birth within the previous six months, or who is breastfeeding.

'Given birth' means 'delivered a living child' or, after 24 weeks of pregnancy, a stillborn child.

As an employer, what are some of the common risks to new or expectant mothers which I need to be aware of?

There are certain aspects of pregnancy which can be exacerbated by various work activities and it is sensible to be aware of these so that special attention can be paid during the Risk Assessment process.

(*a*) Morning sickness and headaches:
- consider early shift patterns
- consider exposure to nauseating smells
- consider exposure to excessive noise.

(*b*) Backache:
- consider excessive standing
- consider posture if prolonged sitting
- consider manual handling tasks.

(*c*) Varicose veins:
- consider standing for prolonged periods
- consider sitting positions and options for foot rests.

(*d*) Haemorrhoids:
- consider working in hot conditions.

(*e*) Frequent visits to the toilet:
- consider location of work area in relation to WCs
- consider ease of leaving the workstation
- consider whether software records absences
- consider confined or restricted working space
- consider ease of leaving the job quickly.

(*f*) Increasing size:
- consider use of protective clothing or uniforms
- consider work in confined spaces
- consider manual handling tasks.

(*g*) Tiredness:
- consider whether overtime is necessary
- consider shift patterns
- consider whether evening work or early morning work is necessary
- consider flexible working hours.

(*h*) Balance:
- consider housekeeping to avoid obstacles
- consider unobstructed work areas
- consider specifically slippery floor surfaces
- consider exposure to wet surfaces.

(*i*) Comfort:
- consider too tight clothing
- consider temperatures of the work area — too cold or too warm
- consider 'over-crowding' with fellow employees
- consider whether tasks need to be done at too great a speed.

(*j*) Stress:
- consider anything which could cause an expectant employee to become anxious about any working conditions.

What steps are involved in completing a Risk Assessment?

As with all Risk Assessment procedures, a planned approach is best.

Risk Assessment is best broken down into steps or stages.

Risk Assessment for new or expectant mothers is best started before any employee becomes pregnant — it is advisable for an employer to assess job activities which *could* cause problems for new or expectant mothers.

Stage 1: Initial Risk Assessment

Take into account any hazards or risks from your work activities (or those of others) which could affect employees of child-bearing age.

Risks include those to the unborn child or to a child of a woman who is breastfeeding.

Look for hazards which could affect all female employees, not just those who are pregnant.

Physical hazards

- Movement and posture
- Manual handling
- Noise
- Shocks or vibrations
- Radiation
- Impact injuries
- Using compressed air tools
- Working underground.

Biological hazards

- Working with micro-organisms which can cause infectious diseases.

Chemicals, gases and vapours

- Toxic chemicals
- Mercury
- Pesticides
- Lead
- Carbon monoxide
- Medicines and drugs
- Veterinary drugs.

Working conditions

- Stress
- Working environment
- Temperatures
- Ventilation
- Travelling
- Violence
- Passive smoking
- Working with DSE
- Working hours
- Use of personal protective equipment
- Use of work clothes or uniforms

- Lone working
- Working at heights
- Mental and physical fatigue
- Rest rooms, work breaks, etc.

Consider whether any of the above can harm any employee, but particularly new or expectant mothers.

The requirement to control many of the above hazards is contained in specific health and safety regulations, e.g. Manual Handling Operation Regulations 1992.

Decide who is likely to be harmed, how, when and how often, etc. Remember that new or expectant mothers may be less tolerant to hazards than other workers and so the degree of control needed to eliminate or reduce the risks may be greater than would normally be expected.

Consult your employees and inform them of any risks identified by the Risk Assessment. In particular, advise all female employees of child-bearing age of any risks which may affect them. Advise them of what steps are to be taken to reduce the risks.

An employee has notified her employer that she is pregnant. What should the employer do?

The employer must carry out a 'specific risk assessment' for the employee concerned. This must take into account any findings from any generic Risk Assessment and must consider any information available regarding any medical condition of the employee.

A 'competent person' must conduct the Risk Assessment and it is best to involve the employee concerned.

Risk Assessments will need to be reviewed regularly and it would be good practice to carry out a new Risk Assessment for each trimester, i.e. at three-monthly intervals as the physical condition of the employee may change and her susceptibility to working conditions may alter, e.g. manual handling activities may need to be prohibited between 6–9 months of pregnancy whereas some manual handling may have been acceptable up to six months. Hormonal changes in women who are pregnant or who have just given birth can affect ligaments, increasing susceptibility to injury.

Any work activity which is identified as a hazard needs to be altered so that the hazard is eliminated or reduced. If this cannot be achieved, the employer must offer the employee alternative work on the same employment terms and conditions.

If alternative, less hazardous work is not available, the employee must be suspended on full pay until ready to return to work or until alternative work is available.

Is it true that jobs which involve pregnant employees standing for long periods of time put the unborn child at risk?

Continuous standing during the working day may cause dizziness, faintness and fatigue. There is an increased risk of premature childbirth and miscarriage.

It would be sensible to review the amount of continuous standing a pregnant employee has to do and seeks ways to reduce it.

Suitable controls will be:

- provide a seat, stool, etc.
- allow the employee to sit
- increase rest periods where there is an opportunity to sit down
- limit formally the amount of time spent standing, e.g. maybe to one or two hours at a time.

What are an employers' responsibilities regarding young workers?

Firstly, a young person is anyone who is under the age of 18 years.

A child is anyone who has not yet reached the official school-leaving age, i.e. 16 years.

If young workers are employed as permanent staff they will be covered by the usual health and safety responsibilities. However, employers have to pay more attention to the hazards and risks to which they may be exposed while at work because a young person may not be able to identify hazard risks and consequences as well as an older person.

Health and safety risks have to be assessed in respect of all young workers.

Students on work experience are classed as employees for the purposes of health and safety law and, whether they are on work premises for one day, one month or any given length of time, a Risk Assessment has to be completed in respect of their work activities.

What needs to be considered in the Risk Assessment for young workers?

Firstly, the Risk Assessment must be completed *before* the young person starts work or work experience.

Each young person must be told what the hazards and risks are and must have control measures explained, etc.

The Risk Assessment must:

- consider the fact that young people are inexperienced in work environments
- consider that young people are physically and mentally immature
- consider their lack of knowledge of work procedures
- consider that they are inexperienced in perceiving danger
- consider that their literacy skills may be less than ideal
- consider all the control measures necessary to reduce or eliminate the hazard

- consider that personal protective equipment, etc., if identified as the control measure, may be sized for adults and may not therefore be 'suitable and sufficient'
- be kept up to date
- be relayed and discussed with the young person
- identify training needs for the young person
- consider the tools and equipment they will use — whether there is an age restriction, e.g. on dangerous machines, cleaning, etc.
- consider the layout of the workplace
- consider the environmental hazards
- consider any hazardous substances in use.

Are there any risks which young people cannot legally be exposed to?

Young people under the age of 18 years must not be allowed to do work which:

- cannot be adapted to meet any physical or mental limitations they may have
- exposes them to substances which are toxic or cause cancer
- exposes them to radiation
- involves extreme heat, noise or vibration.

If young people are over the age of 16 years, however, they may be able to undertake or be exposed to the above tasks and hazards if it is necessary for their training and if they are under constant supervision by a competent person.

Children below the school-leaving age must *never* be allowed to undertake the tasks or be exposed to the hazards listed above.

What training and supervision do young people require under health and safety?

Young people need training when they start work — *before* they undertake any work activity, process or task. They must be trained to do the work without putting themselves or others at risk.

It is important to check that young people have understood the training and information they have been given.

- Do they understand the hazards and risks of the workplace?
- Do they understand the basic emergency procedures, e.g. fire evacuation, first aid, accident reporting?
- Do they understand the control measures in place to eliminate or reduce risks?
- Do they understand their responsibilities as employees not to interfere with safety equipment, not to fool about, etc.?

Young people must be regularly supervised by competent people, i.e. those who understand that a young person may inadvertently put themselves or others at risk because they do not know any better or cannot perceive the risk.

20

Work equipment

What type of equipment is covered by the Provision and Use of Work Equipment Regulations 1998 (PUWER)?

Generally, any equipment which is used by an employee while at work is covered by the Regulations.

Examples of work equipment include:

- photocopiers and printers
- ladders and access towers
- electrical appliances, e.g. kettles
- knives
- hand tools
- power presses
- drilling and sawing machines
- lifting equipment
- motor vehicles
- pressure cleaners
- industrial robots
- industrial machinery.

If employees bring their own equipment into work to use, e.g. maintenance tools, hairdressing equipment, then the employee's own equipment is classed as being work equipment and the employer is responsible for ensuring that it complies with the Regulations.

The Regulations now apply to all work equipment, including mobile work equipment which had to comply fully by December 2002.

Uses of equipment include:

- stopping
- starting
- repairing
- modifying
- maintaining
- servicing
- cleaning
- transporting.

So, maintenance engineers in a factory who only ever repair the equipment but never actually use it on the production line will nevertheless be classed as a user of the equipment.

What equipment is not covered by the Regulations?

The Provision and Use of Work Equipment Regulations 1998 do not apply to equipment used by members of the public, e.g.:

- compressed air equipment in a garage
- amusement machines
- vending machines.

However, the general provision of the Health and Safety at Work Etc. Act 1974 applies and employers have to consider the hazards and risks associated with their undertaking in respect of persons who are not their employees.

Do employees have duties under the Regulations?

No, employees do not have any specific duties under PUWER.

But employees have duties under Sections 7 and 8 of the Health and Safety at Work Etc. Act 1974 not to interfere with equipment given to them by their employer for use at work, nor to engage in any reckless acts, etc.

What do the PUWER Regulations require an employer to do?

Employers must comply with PUWER and, in particular, must ensure that work equipment is:

- suitable for use
- suitable for the purpose and conditions in which it is to be used
- maintained in a safe condition for use so that people's health and safety is not put at risk
- inspected in certain circumstances to ensure that it is and continues to be, safe to use
- inspected by competent persons
- subject to written records of inspection and maintenance.

Employers must also ensure that all employees and others, e.g. contractors, using work equipment have had suitable information, instruction and training.

Risks created by the use of work equipment must be eliminated or controlled so that hazards and risks are minimal.

What risks arise from the use of work equipment?

Many things can cause a risk, for example:

- using the wrong equipment for the job, e.g. ladders instead of access towers for work at high levels

- lack of guards or poor guards on machinery, leading to accidents caused by entanglement, shearing, cutting, trapping, etc.
- having inadequate controls or the wrong type of control so that equipment cannot be turned off quickly and safely or it starts accidentally
- failure to keep guards, safety devices, controls, etc. properly maintained so that machines or equipment become unsafe
- failure to provide the right information, instruction and training for those using the equipment.

When identifying the risks, think about:

- all the work that has to be done with the machine and other equipment during normal usage and during setting up, maintenance, repair, breakdowns and removal of blockages
- who uses the equipment, including experienced and well trained employees and also new starters and those who may have particular difficulties, e.g. impaired mobility
- employees who may act foolishly or carelessly or make mistakes
- whether guards or safety devices are poorly designed and inconvenient to use or easily defeated as this may encourage employees to risk injury
- the type of power supply, e.g. electrical, hydraulic or pneumatic — each have different risks.

What controls can be implemented when using work equipment?

Hazards can be eliminated or controlled by taking a number of measures in relation either to the machine itself, or by following a safe system of work.

Alterations or controls which affect the machine or equipment itself are usually referred to as 'hardware measures'.

Controls which rely on the way people do things are called 'software measures'.

What are some of the 'hardware measures' that can be considered for work equipment?

Guarding

Controlling the risk often means guarding the parts of the machines and equipment that could cause injury.

- Fixed guards should be used wherever possible and should be properly fastened in place with screws or nuts and bolts which need tools to remove them.
- If employees need regular access to parts of machines and a fixed guard is not possible, use an interlocked guard for those parts. This will ensure that the machine cannot start before the guard is closed and will stop if the guard is opened while the machine is on.
- In some cases (e.g. on guillotines, etc.), devices such as photo-electric systems or automatic guards may be used instead of fixed or interlocked guards.
- Check that guards are convenient to use and not easy to defeat, otherwise they may need to be modified.
- Think about the best material for guards — plastic may be easy to see through but can be easily scratched or damaged. If wire mesh or similar materials are used, make sure the holes are not large enough to allow access to the danger area. As well as preventing such access, a guard may be used to prevent harmful fluids, dust, etc. from coming out.
- Make sure that the guards allow the machine to be cleaned safely.

- Where guards cannot give full protection, use jigs, holders, push sticks, etc. to move the work piece.

Selection and siting of controls

Some risks can be reduced by careful selection and siting of the controls for the machinery and equipment, for example:

- position 'hold to run' and/or two-hand controls at a safe distance from the danger area
- ensure that control switches are clearly marked to show what they do
- make sure that operating controls are designed and placed to avoid accidental operation, e.g. by shrouding start buttons and pedals
- interlocked or trapped key systems for guards may be necessary to prevent operators and maintenance workers from entering the danger areas before the machine has stopped
- where appropriate, have emergency stop controls within easy reach, particularly on larger machines so they can be operated quickly in an emergency.

Before fitting emergency stop controls to machines that have not previously had them fitted, it is essential to check that they themselves will not become a risk. For example, some machines need the power supply to be on to operate the brakes. This power could be lost if the machine is stopped using the emergency stop control.

What are some of the 'software' controls available to employers?

Use and maintenance of hand tools

Many risks can be controlled by ensuring that hand tools are properly used and maintained; some examples are given below.

- Hammers: avoid split, broken or loose shafts and worn or chipped heads. Heads should be well secured to the shafts.
- Files: these should have a proper handle and should never be used as levellers.
- Chisels: the cutting edge should be sharpened to the correct angle. Do not allow the head of chisels to spread to a mushroom shape — grind off the sides regularly.
- Screwdrivers: never use these as chisels and never use hammers on them. Split handles are dangerous.
- Spanners: avoid splayed jaws. Scrap any which show signs of slipping. Have enough spanners of the right size. Do not improvise by using pipes, etc. as extension handles.

Maintenance procedures

Make sure that machinery and equipment are maintained in a safe condition. Controlling the risk often means carrying out preventative checks and maintenance.

- Check what the manufacturer's instructions say about maintenance to ensure that this is carried out where necessary.
- Routine daily and weekly checks may be necessary, e.g. fluid levels, pressures, brake function. When you enter into a contract to hire equipment, particularly a long-term one, you need to establish what routine maintenance is needed and who will do this.
- Some equipment, e.g. cranes, needs preventative maintenance, i.e. servicing, so that it does not break down.
- Equipment such as lifts, cranes, pressure systems and power presses should have a thorough examination by a competent person at intervals in law.
- Make sure that the guards and other safety devices are routinely checked and kept in working order. They should also be checked after any repairs or modifications.

Carry out maintenance work safely. Many accidents occur during maintenance work. Controlling the risk means following safe working practices, as detailed below.

- Maintenance should be carried out, where possible, with the power to the equipment off and ideally disconnected or with fuses or keys removed, particularly where access to dangerous parts will be needed.
- Isolate equipment and pipelines containing pressurised fluid, gas, steam or hazardous material. Isolating valves should be locked off and the system depressurised where possible, particularly if access to dangerous parts will be needed.
- Support parts of equipment that could fall.
- Allow moving machines to stop.
- Allow components which operate at high temperatures to cool.
- Switch off the engine of mobile equipment, put the gearbox in neutral, apply the break and, where necessary, check the wheels.
- To prevent fire or explosions, thoroughly clean vessels that have contained flammable solids, liquids, gases or dust and check them before Hot Work is carried out. Even small amounts of flammable material can give off enough vapour to create an explosive air mixture which could be ignited by a hand lamp or welding torch.

Information, instruction and training

Instruct and train employees. Make sure that employees have the knowledge they need to use and maintain equipment safely, for example:

- give them the information they need, e.g. manufacturer's instructions, operating manuals

- instruct them on how to avoid risks, e.g. checking that the drive mechanism is not engaged before starting the engine or machine and do not use on sloping ground
- an inexperienced employee may need some instruction on how to use hand tools safely
- as well as instruction, appropriate training will often be necessary, particularly if control of the risk depends on how an employee uses the work equipment.

Training may be needed for existing staff as well as inexperienced staff or new starters (do not forget temporary staff), particularly if they have to use powered machinery. The greater the danger, the better the training needs to be. For some high risk work, such as driving fork-lift trucks, using a chainsaw and operating a crane, training is usually carried out by specialist instructors. Remember, younger people can be quite skilful when moving and handling powered equipment, but they may lack experience and judgement and may require supervision initially.

People who carry out servicing and repairs should have enough knowledge and training to enable them to follow safe working practices. People under the age of 18 cannot clean any machinery if the act of cleaning exposes them to any risk.

What should employees do to ensure the safe use of equipment?

They should check that:

- they know how to use the machine
- they know how to stop the machine before they start it
- all guards are in position and all protective devices are working
- the area around the machine is clean, tidy and free from obstruction
- they are wearing appropriate clothing and equipment, such as safety goggles or shoes where necessary.

Case studies

Some examples of equipment which commonly is misused and the cause of many accidents.

Ladders

Accidents can happen due to:

- ladders not being securely placed and fixed
- climbing with loads
- overreaching and overbalancing
- ladders being used when other equipment would be safer
- the use of poorly maintained and/or faulty ladders.

Food processing machinery

Accidents can happen due to:

- fingers coming into contact with rotating blades, cutters or knives
- contact with rollers
- contact with feed mechanisms.

Many injuries are caused when well-intentioned operators or service workers remove guards and try to clear blockages with the power switched on. They should switch the power off first. Employees should be trained to follow laid down procedures and safe systems of work, developed for operators and maintenance workers.

Pressure water cleaners

Accidents can happen due to:

- electric shock (often fatal)
- fluid injected through the skin.

The weakest parts of the cleaners are their cables. So, wherever possible, cleaners should be fixed in place and permanently connected to the electrical system. Electrical faults to the plug, cable or equipment may make the metal lance at the end of the flexible hose, or the machine's casing, live. Contact will result in an electric shock.

Machines should be given a regular visual examination, looking for signs of faults or damage and should be checked by the user before use. Faulty or damaged machines must be repaired before use. A *residual current device* should be used in the electricity supply to any cleaner that is not fixed in place.

High pressure jets can force fluid into the skin or eyes. This can be very dangerous, so suitable eye protection and special clothing may be needed.

They should:

- tell the supervisor at once if they think a machine is dangerous because it is not working properly or any guards or protective devices are faulty
- stop using the machine until the matter has been checked.

They should *never*:

- use a machine unless they are trained to do so
- try to clean a moving machine if this could be dangerous — they should switch it off and unplug or lock it off
- use a machine or appliance which has a danger sign or tag attached to it. Danger signs should be removed only by an authorised person who is satisfied that the machine or process is safe
- wear dangling chains, loose clothing, gloves or rings, or have long hair which could get caught up in moving parts
- distract other people who are using machines, fool around or deliberately misuse the equipment
- allow passengers to be carried on vehicles such as dumper trucks or fork-lift trucks unless the vehicle is designed for such use.

What sort of training will Enforcement Officers expect to see an employer undertake?

The Health and Safety Commission (the policy-forming arm of the HSE) have issued a strategy for Training in Health and Safety, including their vision:

> Everyone at work should be competent to fulfil their roles in controlling risk.

Health and safety training covers all training and developmental activities aimed at providing workers and managers with:

- greater awareness of health and safety issues
- specific skills in risk assessment and risk management
- skills related to hazards of particular tasks and occupations
- other skills relating to job specification and design, contract management, ergonomics and occupational health.

Enforcement Officers will expect to see documented training records, training plans and schedules which will achieve the objectives set out above.

21

Violence at work

What is violence at work and, as an employer, is it my responsibility?

People who deal directly with the public may face aggressive, violent or verbally abusive behaviour. They may be sworn at, threatened, physically attacked, assaulted or even have their personal possessions damaged, e.g. cars vandalised, etc.

Violence in respect of work-related incidents is:

> any incident in which a person is abused, threatened or assaulted in circumstances relating to their work.

Verbal abuse and threats used to be the most common forms of violence but, unfortunately, physical abuse and assault is becoming increasingly common.

As an employer, any situation which your employee finds themselves in is your responsibility if it is in the course of their work.

Health and safety regulations require you to complete a Risk Assessment for the likelihood of employees facing violent and aggressive situations.

Are there some employees who are more at risk than others?

The majority of employees who are at risk of violence while they are at work are:

- employed in the Health Service
- teachers
- representing authority
- security guards or bouncers, etc.
- carrying out delivery and collection
- representing public authorities, e.g. Social Services
- police, fire and ambulance personnel.

One survey conducted by one of the teaching unions has confirmed that a teacher faces abuse every seven minutes.

The National Health Service is embarking on a 'zero tolerance' campaign for workplace violence because the incident rate has risen alarmingly.

In the retail sector, it is alleged that over 20 000 workers were assaulted physically in 2001. Verbal abuse is a daily occurrence.

What are the legal requirements in respect of managing the risks of violence to employees?

Section 2 of the Health and Safety at Work Etc. Act 1974 requires all employers to ensure, so far as is reasonably practicable, the health, safety and welfare of their employees.

The Management of Health and Safety at Work Regulations 1999 place more specific requirements on employers — to assess the risks, prevent or control them and develop a clear plan to achieve this.

Findings should be recorded in such a way that employees as well as employers can understand what action they need to take.

Under the Reporting of Injuries, Diseases and Dangerous Occurrences Regulations 1995, any major injury or 'over three day' injury must be reported to the enforcing authority if it has been caused to an employee in the course of their duties while at work.

Employers must consult with their employees and Trade Union Safety Representatives on matters relating to their health and safety, including issues regarding violence at work.

As an employer, what do I need to do to manage violence in the workplace?

The Health and Safety Executive advocate a *four-stage* approach, namely:

Stage 1: Find out if you have a problem/Identify the hazards
Stage 2: Decide what action to take
Stage 3: Take action
Stage 4: Check what you have done.

The four-stage approach is continuous as violence in the workplace may constantly be changing.

Stages 1 and 2 really require a Risk Assessment to assess the following.

- What are the hazards?
- Who might be harmed?
- What are the risks of being harmed?
- What are the consequences?
- What needs to be done to eliminate or control the risks?

Stage 1: Find out if you have a problem/Identify the hazards

- Ask your employees what they think the threat of violence or abuse is in the workplace.
- Create open forums to discuss workplace issues.
- Listen to concerns.

- Train managers to be aware of the likelihood of violence or abuse.
- Check out any bullying — it, too, is a form of violence and abuse.
- Conduct a survey via questionnaires.
- Keep detailed records of all incidents, including verbal abuse, threats, acts of physical violence.
- Create a supportive atmosphere so that employees feel comfortable about reporting incidents of verbal or physical abuse.
- Review accident/incident records.
- Analyse all recorded incidents for trends, patterns, etc. Is it one particular customer or does only one member of staff seem to be affected?
- Predict what might happen — apply a little foresight — if you change opening times for the help desk it may be obvious that customers will get annoyed, etc.

Stage 2: Decide what action to take

- Who might be harmed and how?
- Why?
- Identify any violent customers, tenants, residents, clients, etc.
- What situations will put employees at greatest risk — face-to-face interviews, visits to people's own homes, etc.?
- Evaluate the risks.
- What control measures are already in place — are they adequate, what more could be done?
- Could the environment be improved for both employee and customer?
- Could better training be given?
- Could information be readily available?
- Could the job be done differently?
- Could security devices be installed, e.g. video surveillance cameras, etc.?

- Could double entry doors and coded locks be installed?
- Could panic alarms and buttons be installed?
- Should two or more people deal with specific situations?
- Could security checks be carried out on individuals?
- Should an employee counselling service be introduced?

Stage 3: Take action

- Record all findings.
- Record all actions taken.
- Implement change.
- Acknowledge the hazards and the psychological consequences.
- Develop a safety procedure and incorporate it into the Safety Policy.
- Train employees in any new procedures.

Stage 4: Check what you have done

- Review procedures regularly.
- Adapt processes to reflect the changes needed.
- Set up a working party to consider the consequences of violence.
- Monitor Risk Assessments regularly.
- Update records.
- Review accident and incident records and absence records.

Finally, think about the victims of violence and take action to avoid any long-term psychological damage to their mental well-being.

Consider:

- debriefing and counselling sessions
- paid time off work

- legal help for employees to sue the perpetrators
- the effect of violence on other employees, e.g. those who witnessed the event
- help employees apply for compensation if appropriate
- seek advice from the National Victim Support schemes.

Remember, also, that the *fear* of violence or abuse is equally stressful to employees as actual abuse. If someone is always worried that they *will* be attacked at some time, they are being put under an unacceptable mental strain which will affect their well-being. Employers are responsible.

Top tips

- Assess the likelihood of physical or verbal abuse.
- 'Discretion is the better part of valour' — change the way you do things so that there is less confrontation.
- Undertake Risk Assessments.
- Offer *bespoke* training to employees.
- Support those who have suffered.
- Keep records.
- Regularly review incidents.
- Introduce a *zero tolerance* approach.

22

Stress

What is stress?

Stress can mean different things to different people and is often a generic term for a set of symptoms which can be related to life, work and environmental situations.

Research has shown that there is a clear link between poor work organisation and subsequent ill-health.

The Health and Safety Executive (HSE) define stress as:

> the adverse reaction people have to excessive pressure or other types of demand placed on them.

Pressure in itself is not necessarily bad and some people thrive on it — it is when the pressure is considered excessive, or never-ending, by the individual that ill-health can result.

Stress is now considered one of the top causes of occupational ill-health in the UK.

Recent HSE research has shown that up to 5 million people in the UK feel 'very' or 'extremely' stressed by their work and approximately half a million people believe they are experiencing stress to such an extent that they are being made ill.

Work-related stress costs society between £3.7 billion and £3.8 billion every year.

What are the effects of stress on individuals?

Every individual is likely to react differently to stress and physical and emotional symptoms will vary. There are, however, some common symptoms which can be used as indicators that an individual is suffering from stress:

- distress and irritability
- lack of concentration
- inability to relax
- inability to think clearly
- difficulty in making decisions
- self-doubt, lack of self-esteem
- less enjoyment in work
- tiredness
- depression
- headaches and muscular pains
- anxiety
- sleeplessness
- skin rashes, loss of weight or loss of hair
- psychiatric illness
- nervous breakdown
- heart disease
- digestive disorders
- upper limb disorders
- poor immune systems
- stuttering or stammering.

The symptoms could be extremely varied and, in themselves, would not necessarily be severe. However, collectively a range of symptoms, coupled with dissatisfaction at work, would be indicative of stress.

What are the effects of stress on organisations?

If employees are suffering from stress while at work, there are tangible costs to the business and the performance of the organisation may be at risk.

Work stress may result in:

- reduced overall performance and productivity of departments, teams, individuals, etc.
- an increase in absenteeism
- an increase in accidents, lost time, stoppages, near misses, etc.
- increased staff turnover
- increased customer complaints for poor service, rudeness, etc.
- an increase in civil claims leading to increased compensation and increased insurance premiums, etc.

What is the law on stress at work?

Work stress is not mentioned in any health and safety legislation, although the HSE do recognise it as a serious work hazard and issue Guidance to Employers on how to deal with it.

The Health and Safety at Work Etc. Act 1974 places a general duty on employers to consider the health, safety and welfare of their employees. The Management of Health and Safety at Work Regulations 1999 place duties on employers to carry out Risk Assessments on all work activities which may subject their employees to hazard and risk of injury or ill-health.

Enforcing authorities expect employers, at the very least, to have completed Risk Assessments on the likelihood of stress in the workplace. If stress is likely and the employer has more than five employees, these Risk Assessments must be in writing.

If Enforcement Officers, i.e. HSE Inspectors or EHOs, believe that there is a risk of stress within the workplace, they could deal with the matter by the service of an Improvement Notice, requiring the employer to take such steps as are listed to reduce stress levels to employees.

The HSE have not yet introduced an Approved Code of Practice on stress but they have issued good practice guides and have set up a section of their website specifically for stress-related advice and information (www.hse/gov.uk/stress). Consideration is being given to the introduction of regulations to cover stress at work but, at present, because the effects of stress are so variable and people's tolerances are so great, it is difficult to set down legal standards as to what is and is not acceptable in respect of employer responsibilities.

The increasing expectation of stakeholders in a business that the business will be responsible and address all social, corporate and financial risks is changing perspectives to stress management. Corporate social responsibility is requiring businesses to have transparent policies on a wide range of issues and, increasingly, businesses are expected to report on their positions in annual reports, etc.

Surely, as an employer, stress is not my concern?

Yes, it is. As stated above, employers have duties to ensure the health, safety and welfare of their employees. Stress should be every employer's concern not only because of the legal responsibilities, but also because of the moral and common-law responsibilities on employers to have a 'duty of care' to their employees.

Employers are at increasing risk of being sued by their employees for stress and many cases have been successful in the civil courts. Employers have a 'duty of care' under common law and, where employees believe that their health has been affected by stress, they could sue. It will be the responsibility of the employer to show that he discharged his duty of care to his employee.

In any civil claim, forseeability is a key factor, i.e. did the employer know that there was a likelihood of stress in the workplace, had anything happened which could lead him to believe that there was a problem or that an individual was at risk.

Civil claims for stress are increasingly successful and stress management is becoming almost a prerequisite for insurance cover under Employers' Liability insurance.

Are there any other factors, other than foreseeability, which must be taken into account by the civil courts before a case can be successful?

Yes. Although cases are increasing, and awards can be high, the Court of Appeal in 2002 set down standards which must be considered for a stress-related case to be successful.

- An employer is only in breach of their duty of care if they fail to take reasonable steps to prevent the employee suffering as a result of stress.
- Indications of impending harm to health must be plain enough for any reasonable employer to realise.
- The size and resources of the employer must be considered in respect of what is reasonable to reduce the risk of stress within the organisation.
- Claimants must show that their injury is a result of their employer's actions in failing to protect their health.
- Employers will not be in breach of their duty of care if they allow a willing employee to continue with their job, despite them stating that they have been suffering stress, if the only solution is to dismiss the employee.
- If employers offer a confidential employee counselling service with referral to treatment services they will be unlikely to be found guilty of breach of duty of care.

Employees must let their employer know if they are suffering from stress. Employers are not expected to identify stressed employees unless identified in Risk Assessments.

Case study

An employee of Northumberland County Council won a civil law case against his employer for stress. He was awarded £175 000 compensation.

The employee had suffered one nervous breakdown caused by work-related conditions. He had time off work and, when recovered, returned to the same job. His employers promised him help and support and work-load management. They failed to deliver. He had another nervous breakdown and had to leave work on ill-health grounds. He sued for breach of duty of care and was successful.

What steps can an employer take to establish whether there is a problem of stress in the workplace?

Be open about the fact that you want to investigate and address any stress issues in the workplace.

Consulting with employees is essential and this can be done via trade union representatives, employee representatives or with the entire workforce. Remember to include everyone — it is not only management who can potentially suffer from stress!

Questionnaires could be a good starting point. They could be anonymous so as not to intimidate people. Stress is still considered a 'weakness' to some people and they will fear for their jobs if they think their bosses believe them to have stress.

Check absenteeism — is it high in certain departments? Is there a significant number of short, non-specific periods of absence for certain workers? Is it higher in one department than another?

Is there any issue with lateness, increased disciplinary proceedings? Is staff turnover high? What information comes out of exit interviews.

Is productivity down, quality deteriorating or customer complaints rising?

Talk to staff individually or in a group. Actively seek information.

Ask employees what they would like to happen. Consider work–life balance initiatives, e.g. flexible working.

Involve employees. Tell them what you hope to achieve. Do not promise to change everything overnight — try to tackle major stress factors within the company first.

Check for any bullying, harassment, etc. Many incidents of stress start with employees feeling bullied or harassed and there being no mechanism for their concerns to be dealt with. Perhaps their line manager is the bully or problem?

In effect, you will be carrying out a Risk Assessment.

Write down any findings.

Consider what you can do to improve things.

What are some of the solutions to managing stress in the workplace?

Stress is generally managed in the workplace by a combination of approaches. Management have the greatest opportunity to contribute to a stress-free environment and, often, awareness training for managers is a key tool in preventing or controlling the hazards of stress.

Some solutions are to:

- let employees contribute to planning their own workloads, setting deadlines, etc.
- introduce clear business objectives and expectations
- set and maintain a good work–life balance and discourage long hours or 'presenteeism' in the workplace
- get management to set good examples
- listen to employees, be approachable, create a supportive atmosphere for individuals, foster team working
- train employees how to prioritise tasks
- match individual skills to job tasks
- train employees on new equipment, new software, etc.
- give individuals greater responsibility for their work and planning schedules
- review the work environment and eliminate external stressors such as excessive noise, poor ventilation, lighting glare, threat of violence, etc.
- provide training in self-development and interpersonal skills
- foster good communication. Keep employees aware of what is going on in the business. Help them to feel involved
- consider 'lone workers' and home workers — they may miss human contact and worry about their performance
- introduce appraisals and regular review meetings
- create a supportive environment — stress symptoms are not signs of weakness
- review procedures so as to ensure that issues such as bullying, harassment and discrimination are eliminated and not tolerated

- treat all information from individuals confidentially. Respect people's concerns and confidences
- ensure that managers fulfil the culture of the organisation and that they support their teams, etc.
- have agreements in place for accessing outside help when necessary, e.g. GP practice, counselling services. Allow employees reasonable time off work to seek help to combat the symptoms of stress.

Top tips

- Talk about the problems being experienced by employees — consult and involve employees.
- Listen actively.
- Be open-minded as to what could change — as long as the 'outputs' are achieved, how you get there may not be important.
- Be flexible — consider work–life balance.
- Comply with the Working Time Regulations — do not expect excessive hours.
- Create a supportive company culture.
- Stress is not a weakness — it is a symptom of something which needs to be improved.
- Give employees some controls over their own work patterns.
- Introduce training programmes on new equipment, software, etc.
- Monitor sickness and absence records.
- Generate some *fun* in the workplace!

23

Training in health and safety

What is an employer responsible for in respect of training employees while they are at work?

The Health and Safety at Work Etc. Act 1974, Section 2, sets out the duties of employers as:

> the provision of such information, instruction, training and supervision as is necessary to ensure, so far as is reasonably practicable, the health and safety at work of his employees.

Training should be considered as a risk control measure for hazards identified within the workplace.

The employer can provide his employees with any combination of instruction, information, training and supervision as is appropriate.

The Management of Health and Safety at Work Regulations 1999, Regulation 13, states that an employer must:

- provide training upon recruitment and induction
- provide training whenever an employee is exposed to new or altered risks in respect of people, machinery, processes, materials, etc.
- provide continuous, repeated training so that employees are given information on current best practice

- provide training in methods which are flexible and adaptable and which meets the needs of special groups of workers, e.g. those with disabilities, literacy problems or language problems
- provide training within working hours as a business necessity and without charge to employees.

What are the 'five steps' to information, instruction and training?

The HSE publish many guidance documents on health and safety matters and in the useful *Five Steps* series it covers training as follows:

Step 1: Determine who needs training
Step 2: Decide what training is needed and what the objectives are
Step 3: Decide how the training should be carried out
Step 4: Decide when training should be carried out
Step 5: Check that the training has worked.

Generally, all employers need to have a policy on health and safety training, i.e.:

- What you are going to do?
- When you are going to do it?
- What subjects will be covered?
- Who will do it?
- How often will it be done?
- What assessment tests will be made?

Increasingly, training records are vitally important to prove that, as an employer, you have discharged your duties. If an accident occurs, the first documents which the investigating officer will want to see will probably be:

Case study

A catering employee had her hand crushed between the lid of a container and its frame as the lid crashed down while she was using it. The EHO was planning to prosecute for the accident, i.e. an unsafe system of work because the container lid had not been secured in position. However, the employer was able to demonstrate that the employee had been given training and did know how to secure the lid into the upright position. She chose not to do it through carelessness. The EHO did not prosecute because the employer had suitable and sufficient Risk Assessments and had training records.

- Risk Assessments and
- training records.

Employees who have been able to demonstrate that they have *not* been given adequate training are more likely to be successful in civil claims than those who have been trained.

What are the legal requirements for instruction and training under the various health and safety regulations?

General health and safety

Management of Health and Safety at Work Regulations 1999

Health and Safety training:

- on recruitment
- on being exposed to new or increased risks; and
- repeated as appropriate.

Health and Safety (First Aid) Regulations 1981

First aiders provided under the Regulations must have received training approved by the HSE.

Health and Safety (Safety Signs and Signals) Regulations 1996

Each employee must be given instruction and training on:

- the meaning of safety signs, and
- measures to be taken in connection with safety signs.

Health and Safety (Consultation with Employees) Regulations 1996

The Regulations provide for training for employee representatives in their functions as representatives (as far as is reasonable). You are required to meet the costs of this training, including travel and subsistence and giving time off with pay for training.

Safety Representatives and Safety Committees Regulations 1977

Sufficient time off with pay for safety representatives to receive adequate training in their functions as safety representative.

Control of Substances Hazardous to Health Regulations 2002

These Regulations cover instruction and training in:

- risks created by exposure to substances hazardous to health (e.g. high hazard biological agents) and precautions
- results of any required exposure monitoring
- collective results of any required health surveillance.

Health and Safety (Display Screen Equipment) Regulations 1992

Adequate health and safety training in the use of any workstation to be used.

Noise at Work Regulations 1989

Instruction and training for employees likely to be exposed to daily personal noise levels at 85 dB(A) or above:

- noise exposure: level, risk of damage to hearing and action employees can take to minimise that risk
- personal ear protectors (to be provided by employers): how to get them, where and when they should be worn, how to look after them and how to report defective ear protectors or noise control equipment
- when to seek medical advice on loss of hearing, and
- employees' duties under the Regulations.

Control of Asbestos at Work Regulations 2002

Instruction and training about risks and precautions for:

- employees liable to be exposed to asbestos
- employees who carry out any work connected with your duties under these Regulations.

Control of Lead at Work Regulations 2002

Instruction and training about risks and precautions for:

- employees liable to be exposed to lead, and
- employees who carry out any work connected with your duties under these Regulations.

Ionising Radiations Regulations 1999

Instruction and training to enable employees working with ionising radiation to meet the requirements of the Regulations, e.g. in radiation protection for particular groups of employees classified in the Regulations.

Provision and Use of Work Equipment Regulations 1998

Employees who use work equipment (including hand tools) and those who manage or supervise the use of work equipment need health and safety training in:

- methods which must be used, and
- any risks from use and precautions.

Personal Protective Equipment at Work Regulations 1992

Employees who must be provided with PPE need instruction and training in:

- risks that the PPE will avoid or limit
- the PPE's purpose and the way it must be used
- how to keep the PPE in working order and good repair.

What instruction and training needs to be given on fire safety?

Fire training, to the extent that employees should know what action to take when fire alarms sound, should be given to all employees and should be included in the induction training. Knowledge of particular emergency plans and how to tackle fires with equipment available may be given in specific training at the workplace. At whatever point training is given, the following key points should be covered:

- evacuation plan for the building in case of fire, including assembly points(s)
- how to use fire-fighting appliances available
- how to use other protective equipment, including sprinkler and other protection systems, and the need for fire doors to be unobstructed
- how to raise the alarm and operate the alarm system from call points
- workplace smoking rules
- housekeeping practices which could permit a fire to start and spread if not carried out, e.g. waste disposal, use of ash bins, handling of flammable liquids, etc.

Fire training should be accompanied by practices, including regular fire drills and evacuation procedures. No exceptions should be permitted at these.

What, specifically, must employers provide in the way of information on health and safety issues to employees?

Management of Health and Safety at Work Regulations 1999.

Information on:

- risks to health and safety
- preventative and protective measures

- emergency procedures including evacuation
- specific health and safety risks for temporary employees
- requirements for any health surveillance
- competent persons
- risks created by other employers.

Control of Substances Hazardous to Health Regulations 2002

Information on:

- risks to health created by exposure to substances hazardous to health (including, for example, high hazard biological agents)
- precautions
- results of any required exposure monitoring
- collective results of any required health surveillance
- safety data sheets.

Chemicals (Hazards Information and Packaging for Supply) Regulations 2002

Safety data sheets or the information they contain are to be made available to employees (or to their appointed representatives).

Manual Handling Operations Regulations 1992

Information on:

- the weight of loads for employees undertaking manual handling, and
- the heaviest side of any load whose centre of gravity is not positioned centrally.

Health and Safety (Display Screen Equipment) Regulations 1992

Health and safety information about display screen work for both operators and users (the Regulations define who is an operator and who is a user).

Health and Safety (First Aid) Regulations 1981

Information on first aid arrangements: including facilities, responsible personnel and where first aid equipment is kept.

Health and Safety (Safety Signs and Signals) Regulations 1996

Each employee must be given clear and relevant information on the measures to be taken in connection with safety signs.

Health and Safety Information for Employees Regulations 1989

Information about employees' health, safety and welfare in the form of:

- an approved poster to be displayed where it can be easily read as soon as is reasonably practicable after any employees are taken on, or
- an approved leaflet to be given to employees as soon as practicable after they start.

Health and Safety (Consultation with Employees) Regulations 1996

Necessary information to enable your employees to take part fully in consultation and to understand:

- what the likely risks and hazards arising from their work, or changes to their work, might be
- the measures in place, or to be introduced, to eliminate or reduce them, and
- what employees ought to do when encountering risks and hazards.

Safety Representatives and Safety Committees Regulations 1977

Necessary information to assist the work of safety representatives nominated in writing by a recognised trade union.

Ionising Radiations Regulations 1999

Information:

- to enable employees working with ionising radiations to meet the requirements of the Regulations
- on health hazards for particular employees classified in the Regulations, the precautions to be taken and the importance of complying with medical and technical requirements, and
- for female employees, on the possible hazards to the unborn child and the importance of telling the employer as soon as they find out they are pregnant.

Control of Pesticides Regulations 1986

Information on risks to health from exposure to pesticides and precautions.

Provision and Use of Work Equipment Regulations 1998

Information on:

- conditions and methods of use of work equipment (including hand tools), and
- foreseeable abnormal situations; what to do and lessons learned from previous experience.

Personal Protective Equipment at Work Regulations 1992

Information on:

- risks that the PPE will avoid or limit
- the PPE's purpose and the way it must be used, and
- what your employee needs to do to keep the PPE in working order and good repair.

Control of Asbestos Regulations 2002

Information about risks and precautions for:

- employees liable to be exposed to asbestos
- employees who carry out any work connected with your duties under these Regulations.

Control of Lead at Work Regulations 2002

Information about risks and precautions for:

- employees liable to be exposed to lead, and
- employees who carry out any work connected with your duties under these Regulations.

Noise at Work Regulations 1989

Information on:

- risk of damage to hearing
- what steps are to be taken to minimise risk
- steps the employee must take to obtain personal ear protection
- employees' obligations.

24

Miscellaneous

Consulting employees on health and safety

Why is consultation important?

Consulting employees on health and safety matters can be very important in creating and maintaining a safe and healthy working environment. By consulting employees, an employer should motivate staff and make them aware of health and safety issues. Businesses can become more efficient and the number of accidents and work-related illnesses can be reduced.

By law, employers must consult all of their employees on health and safety matters. Some workers who are self-employed, for example for tax purposes, are classed as employed under health and safety law.

What does consultation on health and safety involve?

Consultation involves employers not only giving information to employees, but also listening to and taking account of what employees say before they make any health and safety decisions.

If a decision involving work equipment, processes or organisation could affect the health and safety of employees, the employer must allow time to give the employees or their representatives, information on what is proposed. The employer must also give the employees or their representatives the chance to express their views. Then the employer must take account of these views before reaching a decision.

What should consultation be about?

Consultation with employees must be carried out on matters to do with health and safety at work, including:

- any change which may substantially affect their health and safety at work, for example in procedures, equipment or ways of working
- the employer's arrangements for getting competent people to help him to satisfy health and safety laws
- the information that employees must be given on likely risks and dangers arising from their work, measures to reduce or eliminate these risks and what they should do if they have to deal with a risk or danger
- the planning of health and safety training
- the health and safety consequences of introducing new technology.

How should consultation take place?

The Safety Representatives and Safety Committee Regulations (SRSCR) 1997

If an employer recognises a trade union and that trade union has appointed, or is about to appoint, safety representatives under the

SRSCR 1997, the employer must consult those safety representatives on matters affecting the group or groups of employees they represent. Members of these groups of employees may include people who are not members of that trade union.

The Health and Safety (Consultation with Employees) Regulations (HSCER) 1996

Any employees not in groups covered by trade union safety representatives must be consulted by their employers under the HSCER 1996. The employer can choose to consult them directly or through elected representatives.

If the employer consults employees directly, he can choose whichever method suits everyone best. If the employer decides to consult his employees through an elected representative, then employees have to elect one or more persons to represent them.

If the employer's arrangements already satisfy the law, there is no need for change.

What help and training will representatives receive?

The employer must make sure that elected representatives receive the training they need to carry out their roles, give them the necessary time off with pay and pay any reasonable costs to do with that training. All representatives must be given time off with pay to take part in any training they need.

All representatives must be given reasonable time off with pay and appropriate help and facilities so they can carry out their role. Candidates for election are also entitled to reasonable time off with pay to carry out their roles.

What information should be available?

Employees or their representatives must be given enough information to allow them to take a full and effective part in the consultation. If the employer so chooses, he can consult both the employees and their representatives about a particular issue.

Employers do not have to provide information that they are not aware of or if it:

- would be against the interests of national security or against the law
- is about someone who has not given their permission for that information to be given out
- would — other than for reasons of its effect on health and safety — harm the business
- if the employer has got the information in connection with legal proceedings.

How are the Regulations regarding employee consultation enforced?

Health and Safety Inspectors enforce the Regulations. If employers do not satisfy the Regulations, they will be committing an offence.

Any employee can apply to an Industrial Tribunal if they feel they have been penalised for taking part in consultation. Representatives who have not received the time off and pay they need to carry out their roles or be trained can also apply.

What do employers need to know about the Pressure Systems Safety Regulations 2000?

If pressure equipment fails in use, it can seriously injure or kill people nearby and cause serious damage to property. Each year in the UK, there are about 150 dangerous occurrences involving such unintentional releases. Around six of these result in a fatality or serious injury.

As an employer or self-employed person, you have a duty to provide a safe workplace and safe work equipment. Designers, manufacturers, suppliers, installers, users and owners also have duties. The main Regulations covering pressure equipment and pressure systems are the Pressure Equipment Regulations 1999 and the Pressure Systems Safety Regulations 2000. Employers have a further duty to *consult any safety or employee representatives* on health and safety matters. Where none are appointed, employers should consult the employees directly.

Examples of pressure systems and equipment include:

- boilers and steam heating systems
- pressurised process plant and piping
- compressed air systems (fixed and portable)
- pressure cookers, autoclaves and retorts
- heat exchangers and refrigeration plant
- valves, steam traps and filters
- pipework and hoses
- pressure gauges and level indicators.

Principal causes of incidents are:

- poor equipment and/or system design
- poor maintenance of equipment
- an unsafe system of work
- operator error, poor training or supervision
- poor installation
- inadequate repairs or modifications.

The main hazards are:

- impact from the blast of an explosion or release of compressed liquid or gas

- impact from parts of equipment that fail or any flying debris
- contact with the released liquid or gas, such as steam
- fire resulting from the escape of flammable liquids or gases.

How do employers reduce the risk of failure from pressure systems?

The level of risk from the failure of pressure systems and equipment depends on a number of factors, including:

- the pressure in the system
- the type of liquid or gas and its properties
- the suitability of the equipment and pipework that contains it
- the age and condition of the equipment
- the complexity and control of its operation
- the prevailing conditions (e.g. a process carried out at high temperature)
- the skill and knowledge of the people who design, manufacture, install, maintain, test and operate the pressure equipment and systems.

To reduce the risks you need to know some precautions, some of which are contained in the Pressure Systems Safety Regulations 2000 and the Pressure Equipment Regulations 1999.

What constitutes safe and suitable equipment?

- When installing new equipment, ensure that it is suitable for its intended purpose and that it is installed correctly. This requirement can normally be met by using the appropriate design, construction and installation standards and/or codes of practice. From 30 May 2002, most pressure equipment placed on the market must meet the requirements of the Pressure Equipment Regulations 1999. For pressure equipment not covered by the Pressure Equipment Regulations 1999, the more general requirements of the Pressure Systems Safety Regulations 2000 apply.

- The pressure system should be designed and manufactured from suitable materials. You should make sure that the vessel, pipes and valves have been made of suitable materials for the liquids or gases they will contain.
- Ensure that the system can be operated safely — without having to climb or struggle through gaps in pipework or structures, for example.
- Be careful when repairing or modifying a pressure system. Following a major repair and/or modification, you may need to have the whole system inspected before it is re-commissioned.

What are ideal operating conditions?

- Know what liquid or gas is being contained, stored or processed (e.g. is it toxic or flammable).
- Know the process conditions, such as the pressures and temperatures.
- Know the safe operating limits of the system and any equipment linked directly to it or affected by it.
- Ensure that there are operating instructions for all the equipment and for the control of the whole system, including emergency instructions.
- Ensure that appropriate employees have access to these instructions and are trained in the operation and use of the equipment or system.

What are protective devices?

- Ensure that suitable protective devices are fitted to the vessels or pipework (e.g. safety valves and any electronic devices which cause shut-down when the pressure, temperature or liquid or gas level exceed permissible limits).

- Ensure that the protective devices have been adjusted to the correct settings.
- If warning devices are fitted, ensure they are noticeable, either by sight or sound.
- Ensure that protective devices are kept in good working condition at all times.
- Ensure that, where fitted, protective devices such as safety valves and bursting discs discharge to a safe place.
- Ensure that, once set, protective devices cannot be altered, except by an authorised person

What needs to be done regarding maintenance?

- All pressure equipment and systems should be properly maintained. There should be a programme of maintenance for the system as a whole. This should take into account the system and equipment age, its uses and the environment.
- Look for tell-tale signs of problems with the system, e.g. if a safety valve repeatedly discharges, this could be an indication that either the system is over-pressurising or the safety valve is not working correctly.
- Look for signs of wear and corrosion.
- Systems should be depressurised before maintenance work is carried out.
- Ensure that there is a safe system of work, so that maintenance work is carried out properly and under suitable supervision.

What training do employees need?

Everybody operating, installing, maintaining, repairing, inspecting and testing pressure equipment should have the necessary skills and knowledge to carry out the job safely — so you need to provide suitable training. This includes all new employees, who should have initial training and be supervised closely.

Additional training or retraining may be required if:

- the job changes
- the equipment or operation changes
- skills have not been used for a while.

What is an examination of equipment scheme?

Under the Pressure Systems Safety Regulations 2000, a written scheme of examination is required for most pressure systems. Exempt systems are listed in the Regulations. Generally speaking, only very small systems are exempt.

- The written scheme should be drawn up (or certified as suitable) by a competent person. It is the duty of the user of an installed system and the owner of a mobile system to ensure that the scheme has been drawn up. You must not allow your pressure system to be operated (or hired out) until you have a written scheme of examination and ensured that the system has been examined.
- The written scheme of examination must cover all protective devices. It must also include every pressure vessel and those parts of pipelines and pipework which, if they fail, may give rise to danger.
- The written scheme must specify the nature and frequency of examination and include any special measures that may be needed to prepare a system for a safe examination.
- The pressure system must be examined in accordance with the written scheme by a competent person.
- For heat generating pressure systems, such as steam boilers, the written scheme should include an examination of the

system when it is cold and stripped down and when it is running under normal conditions.

The key steps are given below.

- Decide what items of equipment and parts of the plant should be included in the scheme. This must include all protective devices. It must also include pressure vessels and parts of pipework which, if they fail, could give rise to danger.
- The scheme must be drawn up (or certified as suitable) by a competent person in accordance with that scheme.

Remember that an examination undertaken in accordance with a written scheme of examination is like an MOT for a car. It is a statutory examination that is designed to ensure that your pressure system is 'roadworthy'. It is not a substitute for regular, routine maintenance.

Who are competent persons under the Regulations?

- You must assure yourself that the competent person has the necessary knowledge, experience and independence to undertake the functions required of them.
- The competent person carrying out the examinations under a written scheme need not necessarily be the same one who prepared or certifies the scheme as suitable.

A competent person may be:

- a company's own in-house inspection department
- an individual person (e.g. a self-employed person), or
- an organisation providing independent inspection services.

Bodies that have United Kingdom Accreditation Service accreditation to the British and European standard BS EN 45004:

1995, for the scope of in-service inspection of pressure equipment, can provide competent persons meeting the appropriate criteria.

The competent person undertaking an examination of a pressure system in accordance with the written scheme of examination takes the responsibility for all aspects of the examination. For example, on systems where ancillary examination techniques (e.g. non-destructive testing) are undertaken, the competent person must assume responsibility for the results of these tests and their interpretation even though the tests may have been carried out by someone else.

Alphabetical list of questions